Mangroves of Andaman & Nicobar Islands : Taxonomy and Ecology

(A COMMUNITY PROFILE)

H.S. Debnath

BISHEN SINGH MAHENDRA PAL SINGH
23-A, New Connaught Place
Dehra Dun - 248 001 (INDIA)

2004

Mangroves of Andaman & Nicobar Islands : Taxonomy and Ecology
(A Community Profile)

ISBN: 81-211-0420-3

Printed by Gajendra Singh Gahlot at Shiva Offset Press, and composed by Doon Phototype Printers, 14 Old Connaught Place, Dehra Dun, India for Bishen Singh Mahendra Pal Singh, Dehra Dun, India.

PREFACE

The mangrove ecosystem of Andaman & Nicobar Islands is one of the most biologically productive, taxonomically diverse and aesthetically celebrated ecosystem of the tropics.

This profile of the mangrove community of Andaman & Nicobar Islands is one in a series of community profiles which treat coastal and marine habitats important to man.

The obvious work that mangrove communities do for man includes the stabilization and protection of shore line; the creation and maintenance of habitats for a great number of animals, many of which are either endangered or have commercial value, the provision of the basis of a food web whose final products include the nutrient of oysters, crabs, lobsters, shrimps and fishes.

The information in this book can give basic understanding of the mangrove community and its role in the regional ecosystem of Andaman & Nicobar Islands.

References are provided for those seeking in-depth treatment of a specific facet of mangrove ecology.

The format, style and level of presentation make this synthesis report adaptable to diversity needs such as the formulation of suitable management plan and the development of a sense of the importance of the resources to those citizens who control this fate.

In this account a descriptive overview of mangrove plants is presented.

Chapter 1 defines mangroves and refers to associated communities; briefly summarizes general features of Andaman & Nicobar Islands; physical conditions of mangrove formations and the community types of the mangroves; extent and distribution of the mangroves in Andaman & Nicobar Islands and considers some of the physiological specializations that allow mangroves to grow in sea water subject to tidal influences and usually in unstable and anaerobic substances. The adaptations necessary for mangrove plants to become establish in and occupy the tidal habitat are varied. In the Chapter 2 autecology of mangroves is described. Chapter 3, the large part of this book, is a systematic survey of mangrove plants and mangroves associates by family. It will serve as a universally applicable florula for students and research workers. Chapter 4 deals with the market valuation and location of the mangrove goods and services. There is a strong commercial significance to the study of mangroves because of their economic value. For example, commercial fisheries are ultimately dependent on mangrove which initiates the food chain of the ecosystem, that aims in a marketable product. Chapter 5 deals with connections between mangroves, sea grass and coral reef ecosystems. Chapter 6 discussed about impact of climatic change and sea level rise on mangroves. The conservation and management of mangrove has to be considered in a special case different from most other ecosystems. Therefore Chapter 7 is thoroughly discussed on the management implications with special reference to the Andaman & Nicobar Islands.

This book has been written for all those with an interest in managing mangrove ecosystems; those with statutory responsibilities; commercial interests; those who lives or work on the mangroves; and those who enjoy this mangroves appreciate its values and qualities.

I hope that this book will be very useful for all concern.

CONTENTS

The water "highway" located between mangrove vegetation.
This open space serves as passage for boats and fishing activities.

Patchwork forest types - Mangrove, Moist deciduous, Evergreen.

Heritiera littoralis Dryand

Mangroves destroyed for cattle.

Knee roots of *Bruguiera gymnorrhiza* (Linn.) Savigny

Pneumatophores of *Heritiera littoralis* Dryand

Bruguiera gymnorrhiza (Linn.) Savigny

Mangroves destroyed for fuelwood.

Prop roots of *Rhizophora apiculata* Blume

Erosion - a loosing battle.

In the developed forests of *Rhizophora apiculata* in Bartang Island (Middle Andaman) the trees are more than 10 m high and considered to be about 40 years old.

Knee like Pneumatophores (breathing roots).

Embankments of stilt roots.

Rhizophora mucronata Lamk.

Rhizophora mucronata Lamk.

Unusual viviparous fruit with two hypocotyles in *Rhizophora mucronata* Lamk.

Rhizophora mucronata Lamk.

A flowering & fruiting branch of *Ceriops decandra* (Griff.) Ding Hou.

Vivipary in *Ceriops tagal* (Parr.) C.B. Rob.

Vivipary in *Rhizophora apiculata* Blume and *R. mucronata* Lamk.

Vivipary in *Rhizophora mucronata* Lamk.

Rhizophora apiculata Blume

Vivipary in *Rhizophora apiculata* Blume

Aegiceras corniculatum (Linn.) Blanco

Aegiceras corniculatum (Linn.) Blanco

Bruguiera cylindrica (Linn.) Bl.

Lumnitzera littorea (J.) Voigt with flowers.

Acrostichum aureum Linn.- a mangrove fern.

Acrostichum aureum Linn. - a population.

Bruguiera gymnorrhiza (Linn.) Savigny

Sonneratia caseolaris (L.) Engl. - a flowering branch.

Sonneratia caseolaris (L.) Engl. - a fruiting branch.

Chapter 1

INTRODUCTION

Mangroves represent dominant soft bottom plant communities of the marine-terrestrial transition in tropical and subtropical regions of the World. As a consequence of the range of spatial and temporal variation in physical factors found in such environments the biological communities display remarkable adaptations, which permit them to survive under such harsh environmental conditions. The plant species are members of terrestrial families, which have adaptations to survive under conditions of high salinity, low oxygen and nutrient availability in the soil, wind and wave action and substrate instability.

The majority of the World's mangrove forests are distributed along the tropical and subtropical shorelines of developing countries. In those areas they are extremely important to coastal societies for both livelihood and commercial purposes. Walsh (1974) reported that from 60 to 75 per cent of the coastlines of the Earth's tropical regions are lined with mangroves, so their importance is clear.

Mangroves habitat in the past has been viewed as economically unproductive area. Until very recently, mangrove swamps were seen as mosquito-infested wastelands, unfit for most uses, but convenient sites for open dumps and sewage discharges. As a consequence have been felled to provide timber for fuel wood and areas for mariculture and agriculture.

Valuation of mangrove forests was previously considered only limited functions provided by these ecosystems - primarily the marketable goods and services (e.g. shoreline protection, coastal productivity, coastal aquaculture, habitat for wildlife and recreation) they produce, which is just a fraction of their total goods and services. As a result, total values and benefits have been underestimated and conversion has become the most attractive alternative. However, with increasing knowledge of both the economic and environmental values of mangrove ecosystem it is understood that mangroves yield greater social net benefits if they remain as natural ecosystem.

Mangrove forests are among the most productive ecosystems on the planet. Their daily rate of organic matter fixation (20 g 0.M/m^2 day) is about 70 times the maximum value reported for tropical oceanic waters and 6 times the mean for the rates reported for marine flagellate blooms in neritic waters in the Caribbean (Ryther, 1969; Burkholder *et al*, 1967). Because of this high rate production of organic matter, they are able to sustain important and valuable populations of fish, shellfish, and wildlife and are prime breeding, spawning, feeding and nursery grounds for many species.

Mangrove supplies natural renewable sources such as fuel wood (including charcoal), tannin, dyes, alcohol, poisons, construction materials (timber and thatching), and food (both plant and animal).

The land-ocean linkages of mangrove ecosystems are important as they produce an extensive variety of goods and services, such as coastline stabilization, protection against erosion, and function as traps for sediment for land accretion and pollutants. Mangrove vegetation filters and purifies water. They are important in the preservation of biological diversity for many plants and animals.

But developmental activities, rapid population growth and migration from inland to coastal areas exert increasing pressure on mangrove ecosystems. In India out of 1000 million populations nearly 25% live in the coastal areas. Today there are troubling signs that large mangrove areas in the developing World have surpassed or come close to the point at which productivity falls and mangrove ecological functions collapse. Degradation of mangrove habitats renders them increasingly susceptible to the potentially adverse effects of climatic change and sea level rise. Loss of fringing mangrove exposes the shoreline to potentially increase in erosion, which may be further, increased under scenarios of global climatic change and sea level rise.

1.1 MANGROVE DEFINITION

The term mangrove is derived from Portuguese and English words viz. "Mangue" and "grove" respectively. A mangrove tree or bush is called Mangue in Portuguese whereas the community of mangrove trees or bushes is Mangal.

The mangrove forest ecosystem consists of the intertidal flora and fauna in the tropics and subtropics and dominated by evergreen sclerophyllous broad-leaved trees with stilt roots or pneumatophores and viviparous seedlings (UNESCO, 1973).

The forest covers the flats between mean sea level and extreme high water (Macnae, 1966). Site condition is characterized by the accumulation of loose mud and silt (Womersley, 1983).

The word mangrove expresses two distinctly different concepts. One usage refers to halophytic species (plant growing in saline soil) of trees and shrubs. In this sense, mangrove is a catch all, botanically diverse, non-taxonomic expression given to approximately 12 families and more than 50 species (Chapman, 1970) of tropical trees and shrubs (Waisel, 1972). All these plants are adapted to loose, wet soils, a saline habitat, periodic tidal submergence and usually have degrees of viviparity of propagules. The second usage of the term mangrove encompasses the entire plant community including individual mangrove species. Synonymous terms include tidal forest, tidal swamp forest, mangrove community, mangrove ecosystem, mangal (Macnae, 1968) and mangrove swamp.

For consistency in this publication "Mangrove" will be referred to individual plant, whereas mangrove forest, mangrove swamp, tidal forest and mangal will be referred to the whole community or association dominated by the plants.

1.2 GENERAL FEATURES OF ANDAMAN-NICOBAR ISLANDS

Physical Geography

Andaman and Nicobar are a group of 572 Islands and islets in the Bay of Bengal. They constitute a biogeographically distinct province located between 6°45′ - 13°40′ N latitudes and 92°12′ - 93°57′ E longitudes. The Andamans are located further away from the equator and are located close to Myanmar, with a few exceptions, the bulk of the Islands are contiguous, whereas Nicobars lie close to the equator, and are disjunctive. The group covers an area of 8293 sq. km, which is 0.25% of total landmass of the Indian Union. The contribution of these Islands to the Mangrove vegetation exceeds 22.5%. The rainforests of the Islands constitute 0.3% of the Asian forests and 0.07% of the global tropical rain forests. The 1986 km long coastline of the Andaman-Nicobar Islands represents one fourth of the Indian coastline. The Exclusive Economic Zone (E.E.Z.) is approximately 6,00,000 sq. km which is roughly 30% of that of India.

Topography

The hypothesis is that the Islands are emergent peaks of a submerged mountain related to the Arakan Yoma range.

The geological formation in the Islands is mainly of thick Eocene sediments deposited on pretertiary fine grey sand stone, silt stones and shoals in which the basic and ultra basic igneous rocks are intrusive. The calcareous sand stones and sand rocks occur frequently interspersed with conglomerates and intercalated clays.

Soil

Mangrove soils are generally distinctive. They are poorly drained, saline, anoxic and rich in organic matter. The primary constituents of the soils range from compact fine-grained clay containing little or no organic matter in the more recent deposits at the sea front and riverbanks to comparatively friable brownish-black mud containing sand and a high percentage of organic matter in the middle of Islands and towards the hinterland. In areas subject to daily tidal flushing, mangrove soils are often ill consolidated and semi fluid in nature. Under the anaerobic conditions below the soil surface, sulphate-reducing bacteria produces hydrogen sulphide through reduction of organic sulphate from the soils. Emission of this gas gives mangrove soils their pungent odour. The characteristic gray or black colour of mangrove soils is due to reduction of ferric compounds to various sulphides by the gas.

In general, soils of the territory are acidic in nature. Soil showed a mean value of 4.5 to 6.0 pH, 0.25 to 1.54 per cent for organic carbon and 0.02 to 0.27 per cent for P_2O_5 (Singh and Mongia, 1985).

Climate

Proximity to the equator and to the sea ensures a hot, humid, uniform climate with day temperature around 30 degrees Celsius in Andaman-Nicobar Islands. There is little variation in mean monthly temperature with the values between 15-33 degrees Celsius. The mean relative humidity is about 82% and the average annual rainfall varies between 300-380 cm. The Islands are subjected to the southwest and northeast monsoon, the former from May to September and the later from November to March. The cyclonic winds are also frequent. The months of January to March remain relatively drier. Details of climatological data at Andaman-Nicobar Islands are presented in Table-1.

1.3 PHYSICAL CONDITIONS OF MANGROVE FORMATIONS

Following physical conditions appear to limit the distribution of mangroves and determine the extent of mangrove ecosystem development.

Temperature

Mangroves are limited in distribution on a global scale by temperature. They are tropical species and do not develop satisfactorily in regions where the annual average temperature is below 19 degrees Celsius (Waisel, 1972). Walsh (1974) and Chapman (1975, 1977) suggested that extensive mangrove development was only possible in areas where the average air temperature of the coldest north exceeds 20 degrees Celsius and where the seasonal range does not exceed 10 degrees Celsius or temperature below freezing for any length of time.

Temperature affects photosynthetic activity, water loss and stomata opening, transpiration and salt loss. Consequently since mangroves occur in a high salinity environment in tropical regions they show xeromorphic adaptations. Moore *et al* (1972, 1973) and Clough *et al* (1982) showed that for most mangrove species photosynthesis declines sharply above 35 degrees Celsius with little or none occurring above 40 degrees Celsius.

High water temperature can also be limiting. McMillan (1971) reported that seedlings of *Avicennia* were killed by temperature of 39 to 40 degrees Celsius, although established seedlings and trees are not damaged.

Salinity

Mangroves are facultative halophytes i.e. salt water is not a physical requirement (Bowman, 1917; Egler, 1948). In fact, most mangroves are capable of growing quite well in fresh water (Teas, 1979). It is important to note, however, that mangrove ecosystems do not develop in strictly freshwater environments. This is probable due to the fact that they are not good competitors and the salinity is important in reducing competition from freshwater and terrestrial vascular plant (Kuenzler, 1974).

Table-1: Climatological Data at Andamans-Nicobar Islands from 1980-1990

Climatic condition		Jan.	Feb.	Mar.	Apr.	May	June	July	Aug.	Sept.	Oct.	Nov.	Dec.
Temperature (°C)	Max.	30.1	30.4	32	33	32.8	31.6	30.5	30.1	31	30.8	30.5	29.4
	Min.	15.4	16	17.4	17.3	19.4	19.3	19	19.3	19	18.8	18.8	17
Monthly rainfall (mm)		15	12	6.5	32	245	440	230	560	350	380	208	55
Cloudiness (okta)		¼	¼	½	¾	¾	¾	¾	¾	½	½	¼	¼
No. of days with fog and haze	Haze	4	4	2	–	–	–	–	–	–	–	3	3
	Fog	1	2	–	–	–	–	–	–	–	–	1	1
Visibility (km)		20 →	→	→	→	10 →	→	→	→	15 →	→	→	→
Prevailing wind		NE	NE	NE	E	SW	SW	SW	SW	SW	SE	NE	E
Wind velocity	Mean	←	←	6	→	→ ←	←	←	16	12 ←	←	8 →	→

Presence of fresh water

Mangroves grow best where the fresh water supply is available in adequate amounts. Fresh water flow from upland carries nutrients and silt important for the growth of mangroves. Mangroves are thus most well developed on muddy coastal plains where adequate fresh water supplies from river discharge are available. Rivers are important geomorphic agents that shape the earth's surface and create deltic features over which mangrove develop. Thus mangroves are well developed in the east coast of Andaman Islands. Even in some places like Baratang, Austin and Straight Islands the species *Rhizophora apiculata* and *Bruguiera gymnorrhiza* may attain a girth of 50 cm and grow 50 m in height. It is to be noted that mangroves may develop in areas where there are no permanent river discharges, but their development in these areas is limited. The absence of river discharges can be mitigated by the availability of run off or fresh water upwelling.

Tidal fluctuation

Mangrove forests develop only in shallow water and intertidal areas and are thus strongly influenced by the tides. Whereas according to Odum *et al* (1982) tidal influence is not a direct physiological requirement for mangroves, it is important indirectly in a number of ways.

First of all, tidal stress in combination with salinity helps exclude other competitively superior vascular plants permitting mangroves to exist. Secondly, tides bring salt-water up to estuaries permitting mangroves to become established well in land. Thirdly, tides may transport nutrients and relatively clean water into mangrove ecosystem and export accumulation of organic carbon and reduced sulphur compounds. Fourthly, in areas with high evaporation rates, the action of the tides helps to prevent soil salinities from reaching concentration which might be lethal to mangroves. Fifthly, tides aid in the dispersal of mangrove propagules and detritus.

Perhaps, these are factors, termed tidal subsides by E.P. Odum (1971) that mangrove communities reach their greatest development where there are large tidal fluctuation.

Substrate and wave energy

Mangroves grow best only where there is no significant wave action. This lack of vigorous water motion has in its turn significant effects.

Slow water movement means that fine sediment particles tend to settle out and accumulate on the bottom. The result is an accumulation of mud and hence, the substrate in mangrove swamps is usually mud.

Mangroves have the ability to grow on many types of substrates. They are also developed, although in a scattered manner in sandy or rocky beaches or on old coral reefs covered with a thin sheet of sand or mud. In these localities the height of the stands is low, trees are dwarfed and crooked and the fringe is very narrow and sparse. It is observed in some places of Middle and South Andaman that *Rhizophora apiculata* and *R. mucronata* are sometimes growing together on rocks or old coral reefs by making small isle.

Many mangroves have peculiar prop (stilt) roots. These dense root systems decrease the water movement. As a result very finest particles settle out around the roots of the mangroves, creating ever-accumulating layers of sediment. Once dropped, the sediment is usually not picked up again and transported out. Thus rapid accumulation of sediment takes place within the mangrove association.

A comparison of the mangrove stands that grow in the Nicobar Islands and those that grow in the Andaman Islands makes the difference obvious.

In Nicobars, the mangrove stands are much less extensive and the trees are relatively small. The main factor controlling the growth of mangrove vegetation seems to be the physical condition of the Islands.

Besides being relatively small, the Islands here are mostly rocky surrounded by deep sea and are influenced by strong wave action. Nearly all of the existing tidal flats are in the form of narrow strips. All these factors restrict the development of mangrove forest in this region.

Shelter

Mangrove seedlings and mature plants are vulnerable to uprooting by waves and current scour. They reach the sea only on protected segments of the coast, on the lee of offshore reefs, shoals or other protective structures. They form a rather uniform, evergreen fringe which is most profusely developed in relatively sheltered areas, along estuaries, coastal lagoons and backwaters where the periodic ebb and flow of the tides bring along mixing of fresh water from rains and land drainage with marine coastal waters.

1.4 EXTENT AND DISTRIBUTION

Mangroves dominate approximately 75% of the World's tropical coastline between 25°N and 25°S latitude (Mc Gill, 1959). A SCOR/ UNESCO biosphere inventory of mangrove lands in the late seventies came up with a total figure of some 162,210 sq.km. of mangrove in World (Saenger *et al*, 1983). Although the distribution of mangroves stretches into higher latitudes (to about 40° North and South, where their spread is prevented essentially by ground forest), their main distribution is in the tropical areas.

The total area of the Indian mangrove forests is estimated to be 356,000 hectares, of which nearly 85% of the Indian mangrove forests are occurring in Sunderbans (West Bengal) and the Islands in the Bay of Bengal. These 85% i.e. about 300,000 hectares of mangrove forests are one of the most important mangrove areas in the World (Blasco, 1977). The irregular and deeply indented coastline of these Andaman-Nicobar Islands, results in innumerable creeks, bays, estuaries that facilitate the development of rich and extensive mangrove forests.

The total mangrove area in Andaman and Nicobar Islands has not been assessed correctly, although several attempts have been made earlier to survey the mangrove areas of Andaman and Nicobar Islands. The first estimate of the mangrove covered area was mentioned in Mangrove symposium, held at Calcutta, by Mathuada (1957) who noted that 116,640 hectares are occupied by mangrove forest in Andaman-Nicobar Islands. In the same symposium Waheed Khan (1957) stated that 105,694 hectares area of Andaman-Nicobar Islands is of mangrove forest. Sahni (1957) estimated 117,100 hectares mangrove area of Andaman-Nicobar Islands. Another estimate of mangrove cover in Andaman and Nicobar Islands was made by Sidhu (1963), to be about 115,200 hectares. According to Blasco (1977) the mangrove-covered area in Andaman-Nicobar Islands is about 115,000 hectares. Mall *et al* (1982) stated that nearly 101,943 hectares land is covered with mangrove forest in Andaman Islands. Action plan for conservation of mangroves of 1989 by Forest Department of

Andaman-Nicobar Islands showed 77,769 hectares covered area of mangroves (Table-2).

Table-2: Distribution of Mangrove Forest in Andaman-Nicobar Islands

Place	Forest covered area (ha.)	Mangrove covered area (ha.)	Percentage
North Andaman	2,09,856	27,312.97	13
Middle Andaman	85,712	23,394.55	27.3
South Andaman	1,31,994	12,750.33	9.7
Baratang	64,651	8,025.60	12.4
Little Andaman	70,649	3,414.22	4.8
Nicobar Islands	1,54,207	2,871.08	1.9
Total	7,17,069	77,768.75	10.85

But it is not clear whether the estimated area covers only wooded regions or the total mangrove environment (mangrove marshes) and also it is unclear whether the variation in the different estimates is due to the rapid rate of deforestation or as a result of different techniques used. Recently remote sensing techniques are used in assessing the mangrove area. But the estimates of mangrove cover area are subject to several sources of inaccuracy or confusion. Firstly, if the canopy is dense and obscures treeless patches and channels, total mangrove cover may be overestimated (Ibrahim and Yusoh, 1992). Conversely mangrove cover may be underestimated if the mangrove occurs in either small patches or low density. This is because small patches of mangrove may be beyond the sensor spatial resolution and low-density mangrove may strongly resemble undestroyed vegetation or substrate. Furthermore, tidal height may have a bearing on the interpretation of intertidal mangrove particularly where base substrate may or may not be exposed.

It will not be an exaggeration to assume that mangrove forests are found in most of the Andaman Islands. Of course, the extent depends upon the climatic condition as well as on the physical characteristics of the Islands. In terms of tree size and extent the mangrove formations are better developed in the following areas:

Andamans:

North Andaman
Austin, Straight Islands and Kalighat creek.

Middle Andaman
Bumlangta, Charulangta.

South Andaman
Baratang, Havelock Islands, Wrafter's and South creeks.

Little Andaman
Dugong Creek

Nicobars:

West Katchal Bay, Nancowry Island, Boloo Chengappa Bay, Ganges harbour and the mouths of Jubiliee, Dogmar, Alexandra, Galathea rivers of Great Nicobar Islands.

1.5 MANGROVE COMMUNITY TYPES

Mangrove forest communities exhibit tremendous variation in their structural development. This structural variability is the response of the mangrove species to the sum total of particular environmental factors.

Snedaker and Pool (1973) and Lugo and Snedaker (1974) identified six major community types resulting from different geological and hydrological processes (Fig. 1.). Each of these community types has its own characteristic set of environmental variables, that control mangrove structure include soil type and depth, soil salinity range, nutrient inputs, tidal fluctuations runoff and groundwater periodicity, droughts. Each community type has characteristic variation of litter decomposition, primary production, carbon export and nutrient cycling rates. The description of these community types is as follows:

Riverine mangrove forests

This type of forests develops along the edges of flowing waters such as rivers, estuaries and creeks. The entire forest is usually flushed by daily tides. Water flows and nutrient inputs are high in this forest. Floodwaters bring in silts and mineral nutrients and these are rapidly incorporated into plant tissue. On the periphery of the forest the species of *Rhizophora* predominate. The species *Rhizophora apiculata* and *R. mucronata* develop

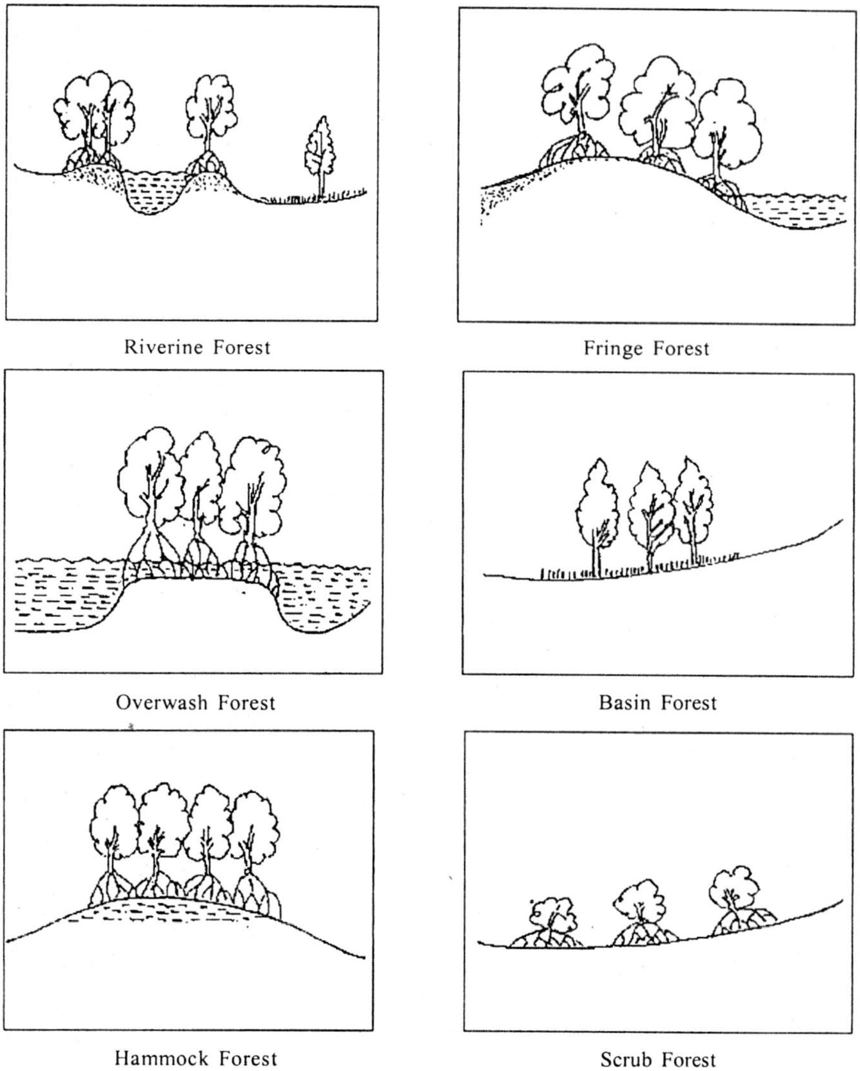

Fig. 1. The six mangrove community types (Lugo and Snedaker, 1974).

a complex maze of prop roots, which allow the establishment of well-developed trees in spite of strong water flows. Inland from the fringe one finds stands *Avicennia* and *Bruguiera* sp. Usually riverine forests are most luxuriant in the lower and middle part of the estuary. Mangroves may reach heights of 25-30 m.

This type of forests is more in Andaman than the Nicobar Islands.

Fringe mangrove forests

Mangroves form a relatively thin fringe along waterways. Fringing mangroves develop along protected shores, over shoals or spits, often forming overwash Islands. Fringes usually have noticeable gradients in topography, turbulence and tidal aptitude. These formations are best defined along shorelines whose elevations are higher than mean high tide. Outer edge of the forests where the levels of kinetic energy are high, the species of *Rhizophora* dominate. In the inner portions *Avicennia* sp. may become dominant. Tree species may reach about 10 m in height.

Overwash mangrove forests

This type of forests is frequently over-washed by tides and thus nutrient inputs are high. All species of mangroves may be present, but members of *Rhizophora* usually dominate. Maximum height of trees is about 7 m.

Basin mangrove forests

These forests occur inland in depressions channeling terrestrial run-off towards the coast. Close to the coast the forests are influenced by daily tides, but the water turnover rate is slow. The structural characteristics of basin forests depend on the hydroperiod. Where water flows are weak but constant, forests develop well. In stagnant basins there may be oxygen depletion, slow nutrient recycling and reduced growth. Basin receives and stores water seasonally. Because of the uniform sheet flows strong salinity gradients do not develop within the basin. Basin forests usually dominated by *Avicennia* sp. Tidal creeks and drainage channels within the basin are often lined with *Rhizophora* sp. Trees may reach up to 15 m in height.

Hammock mangrove forests

This type of mangrove communities is similar to the basin type except that they occur on ground that is slightly elevated to surrounding areas. All species of mangrove may be present.

Scrub or dwarf forests

This community type occurs where growth is limited by edaphic factors. Stands of dwarf *Rhizophora apiculata* and *R. mucronata* develop in some areas of middle and south Andaman over peat substrates in basins that do not receive substantial amounts of terrestrial run-off. Nutrients appear to be limiting although substrate must play a role.

In these areas, trees are dwarfed and crooked and the fringe is very narrow and sparse.

Dwarf *Avicennia* sp. is often found on the landward side of fringes and basins immediately adjacent to salt flats or hyper saline lagoons. The dwarfing factor here is extremely high levels of salt in the soil.

Mangroves are also developed, although in a scattered manner, in sandy or rocky beaches or on old coral reefs covered with a thin sheet of sand or mud. Trees of these localities rarely exceed 1.5 m. Many of these tiny trees are 30 or more years of age. In some places of middle and south Andaman the dwarf *Rhizophora apiculata* and *R. mucronata* are sometimes growing together on rocks or old coral reefs by making a small isle.

1.6 SUBSTRATES

Mangroves are found on a wide variety of substrates, including fine, inorganic muds, muds with a high organic content, peat, sand, reef flat, dead coral and even on rock, if there are sufficient crevices for root attachment. Mangrove ecosystems, however, appear to flourish only on muds and fine grained sands (Odum, 1982).

In Andaman-Nicobar, the mangrove forest soil can be broadly divided into two main categories a) halic hydraquents - unripe clayey soils, permanently saturated with water, with low bulk densities (+0.7), high degree of subsidence after drainage, pH more than 5.5, high base saturation and high cation exchange capacity, b) halic sulfaquents- young clayey soils, permanently saturated with water, having sulfidic material within 50 cm of the mineral soil surface, pH 5.0 and high base saturation (Dagar *et al*, 1991).

Sediment distribution, mangrove development is controlled to a considerable extent by wave and current energy. Low energy shorelines accumulate fine-grained sediments such as mud and silt and usually have the best mangrove growth. Higher energy shorelines (more wave action or higher current velocities) are characterized by sandy sediments and less productive mangroves.

pH value of Andaman-Nicobar mangrove peats is usually about 6. The pH values of the surface soils between the forest margin and about 20 m from the edge tended to be more acidic than those at the open area and the more inland regions (Dagar *et al*, 1991).

Presumably, the acidic character of mangrove peat results from release of organic acids during anaerobic decomposition and from the oxidation of reduced sulphur compounds if the peat is dried in the presence of oxygen.

Dagar *et al* (1989) carried out measurements on more parameters such as nutrient contents, exchangeable cations and also compared the properties of soil collected from the different mangrove communities of Andaman-Nicobar Islands.

1.7 WATER QUALITY

Water quality characteristics of surface waters flowing through Andaman-Nicobar mangrove ecosystems exhibit great variation from one location to the next. Proximity to terrestrial ecosystems, the tidal influences, seasonal rainfall, evaporation and human activities are all important in determining overall water quality.

In general, the surface waters associated with mangroves are characterized by

a) a wide range of salinities from virtually fresh water to about 40 ppt.

b) low macronutrient concentration (particularly phosphorus).

c) relatively low dissolved oxygen concentrations.

d) Frequently increased watercolour and turbidity (Odum, 1982).

Under the influence of rainfall, evaporation and tidal flushing the salinity of Andaman and Nicobar coastline may range from 25-30 ppt., while that of soil water shows a marked fluctuation ranging from 3 to 40 ppt. (Dagar *et al*, 1991).

The surface salinity of the Bay water ranged between 8 ppt. in January and 35.2 ppt. in April.

The surface seawater temperature of Andaman fluctuated between 26°and 33°C, while tidal creeks water shows a fluctuation ranging from 26°to 34°C (Dorairaj and Soundararajan, 1987a).

Chapter 2

AUTECOLOGY OF MANGROVES

2.1 ADAPTATIONS

Adaptability is one of the essential features of living organisms and plants are no exception. They are, therefore, capable of changing gradually in such a way as to withstand or prevent a strain when subjected to a stress, i.e. any environmental factor potentially unfavourable to living organisms.

Adaptations may be either stable, having arisen by evolution over a large number of generations, or unstable, depending on the developmental stage of the plant and the environmental factors to which it has been exposed. The unstable adaptation must, of course, also have arisen by evolution, but the hereditary potential is wide enough to permit large changes during the growth and development of the organisms.

Thus adaptation is important both in the elastic and plastic strains. A strain may be completely reversible. Such reversible strains are said to be elastic. Beyond this point, the strain will be only partially reversible, and the irreversible part is called a plastic strain.

An adapted organism, on the other hand, may live, grow, complete its life cycle, and regenerate in the presence of the stress. This kind of adaptation has been called "capacity adaptation" (Precht, 1967), in which specific morphological and physiological properties enable the plants to increase their capacity to carry out metabolic reactions in potentially unfavourable environmental factors. Another kind of adaptation is "functional adjustment" which relies on a mixture of phenotypic plasticity and genetic variation to allow plant populations to change both morphologically and physiologically in response to oscillations in climate or small changes of micro habitat. The relative importance of phenotypic plasticity (the ability of the individual plant to change during its lifetime) as opposed to genetic variation (heritable variation in populations) varies both between species and within populations of the same species. The crucial difference in terms of plant resources to the environment is that phenotypic plasticity can change the phenotype of the plant during the lifetime of the individual and is therefore a potential immediate reaction to environmental stress, while genetic variation allows change from one generation to another as a response to selection.

Several environmental factors determine the position of individual species in a mangrove forest. For instance, in a fringe one may observe variations in terms of salinity gradients, tidal exposures and different soil types and textures with distance inland. Each species has adaptations that allow it to cope best with the site factors with the least metabolic lost. This means that more energy can be allotted for growth and reproduction allowing that species to become highly competitive and ultimately dominant. This may lead to an observable zonation of species in a mangrove stand.

The major components of mangrove swamps are the salt tolerant trees and shrubs, which are mostly unrelated plant species. They nevertheless have similar physiogamy, physiological characteristics and structural adaptations to the habitat as a result convergent evolution (Yanney-Ewusie, 1980).

Mangroves have a series of remarkable adaptations, which enable them to flourish in an environment characterized by high temperatures, wide fluctuating salinities, and shifting of anaerobic substrates. The ability to exploit the brief availability of mangrove resources depends on capacity adaptations and also functional adjustment. The various tree species differ in their growth pattern and survival and responses to the regular tidal variations and duration of inundation. The mangroves are adapted to high water stress, exhibit mechanisms that permit water uptake against a gradient and have xerophytic adaptations, extensive and specialized root system (Dawes, 1981).

2.1.1 Natural stress - anaerobic sediments

Mangroves are characterized by its remarkable root formations. The root system of mangroves provides the key to existence upon unfriendly substrates. The types of roots are obviously tolerant to withstand the conditions for nutrition, assimilation of water and oxygen in anaerobic mud and also for anchorage on an unstable substratum. Members of the mangrove vegetation have shallow root system. Because the soil is mostly anoxic except for the surface layers where the roots are spread. This anoxic nature of the soil is due to the presence of ferrous sulphides, which are ferric compounds reduced hydrogen sulfide. Unlike higher plants, mangroves usually have highly developed aerial roots and modest below ground root systems. The aerial roots allow atmospheric gases to reach the underground roots, which are embedded in anaerobic soils.

Very complex prop root mazes develop in some fringe forests, possible as a response to wind and wave stresses. The trees growing on more stable substrates often develop very erect boles and have proportionately fewer prop roots. These characteristics, the ability to become implanted in deeper water (up to 30 cm), and to withstand dislodgment by water movement, contribute to allow this species to become dominant in the outer fringe.

In coastal regions of Andaman and Nicobar Islands the species *Rhizophora apiculata* and *R. mucronata* are the pioneers. Stems of *Rhizophora* sp. are provided with a cone of branches stilt or prop roots, which extend a meter or more above the surface of the soil and also forked aerial roots stick out or hang down to reach eventually the soil as they grow towards the sea. These roots contain many small pores (lenticels), which at low tide allow oxygen to diffuse into the plant and down to the underground roots by means of passages called aerenchyma (Scholander *et al*, 1955). The lenticels are highly hydrophobic and prevent water penetration into the aerenchyma system during high tide (Waisel, 1972).

Over a period of many years mangrove trees can stand to hostile swamp environment where the oxygen supply or its availability is minimal. Moreover, the presence of high percentage of salt limits water uptake by plants. The relative stagnation of water in Basin forests requires special adaptations to allow ventilation of the root system. Mangroves have adaptations to overcome this problem. They ventilate their root system by producing the shallow roots often send up extensions to the surface of the substrate, called pneumatophores. The surface of the pneumatophores has minute pores through which gas exchange takes place. In this way oxygen diffuses down to the root and carbon dioxide is vented out.

The members of the genera *Avicennia* and *Sonneratia* do not have prop roots, but do possesses small air roots or pneumatophores, which extend vertically upward from the underground roots and project from the soil and the stagnant water surface to a height of 20 to 30 cm above the soil. These pneumatophores resemble hundreds of tiny fingers sticking up out of the mud underneath the tree canopy. At low tide, air travels through the pneumatophores into the aerenchyma system and then to all living root tissues.

The members of the genus *Xylocarpus* have stout and blunt pneumatophores. The horizontal roots of *X. moluccensis* possess knob producing thick conical pneumatophores and in *X. granatum* these roots have an upper thickening, which appears as a sharp ridge above the soil.

In the members of the genera *Bruguiera* and *Ceriops* lateral roots come above the surface of the soil and make knee like curving thickened bends and return to the soil. These knee roots moving downward into the soil become apogeotropical and produce absorbing roots. Air travels through the lenticels, which are common on the knee roots. The roots of the members of *Lumnitzera* emerge above the soil surface and curve down again but do not form thickened knees.

Mangroves achieve structural stability in at least two ways. Species of the genus *Rhizophora* use the system of prop roots to provide a firm foundation for the tree. The horizontal extent of the prop root system insures considerable protection from all kind of worst condition like wave action, storm, unstable soil etc. Other mangrove species of the genera *Avicennia* and *Sonneratia* obtain stability with an extensive system of shallow, underground "cable" roots that radiate out from the central trunk for a considerable distance in all directions. The pneumatophores extend upward from the cable roots. The cable roots of *Heritiera* and *Xylocarpus* form aerial flangers of planks. These planks like roots of *Heritiera* travel several meters and form several partitions around trunks and help in shore building by keeping the sand intact. From the perspective of salinity, the prop roots of *Rhizophora* appear to offer distinct advantage from the cable roots of *Avicennia* and *Sonneratia* where wave action and current energies are high. But from the standpoint of effectiveness in transporting oxygen to the underground roots, both prop roots and cable roots seem equally effective.

2.1.2 Natural Stress - Salinity

Mangroves accommodate fluctuations and extremes of water and soil salinity through a variety of mechanisms. There are two major methods of internal ion regulation i) the salt exclusion and ii) the salt excretion (Scholander *et al*, 1962).

The salt excluding species, the species of the genus *Rhizophora* separate fresh water at the root surface by means of a non-metabolic ultra filtration system (Scholander, 1968), where the passage of water and essential nutrients is thus allowed but salts are not permitted to enter.

A negative pressure in the xylem, which results from transpiration at the leaf surface, powers this reverse osmosis process. The salt concentration in the sap of salt excluding mangroves is about $1/70^{th}$ the salt concentration in seawater, although this concentration is almost 10 times higher than the normal plants (Scholander *et al*, 1962).

The salt secreting species, use salt glands on the leaf surface to excrete excess salt. This strategy is best exemplified by the species of *Avicennia*, where salt is freely allowed to enter the xylem but it passes out through the salt glands and crystallizes on the leaf surface. This process appears to involve active transport with a requirement for biochemical energy input. Salt concentration in the sap of salt secreting species is about $1/7^{th}$ concentration of seawater, which is almost 10 times higher than that of the salt excluding species (Atkinson *et al*, 1967).

Besides these two general mechanisms, individual species utilize a variety of mechanisms to maintain suitable salt balance (Albert, 1975). Members of the genus *Rhizophora* store and ultimately dispose off excess salt in leaves and fruits (Teas, 1979). Most salt secretors are capable of limited salt exclusion at the root surface. Another characteristic feature of mangroves is thick succulent leaves. These succulent leaves enable the plant to withdraw water stress and through the transpiration process excess salt is released as epidermal or laminal secretions, which in turn are washed by the rainwater or evaporated into the humid atmosphere.

Each mangrove species is associated with an optimum salinity, which varies from species to species (Snedaker, 1978). Each species occupies the salinity zone to which it is best adapted physiologically as compared to potential competition species. The members of the genus *Rhizophora* are probably limited by soil salinities above 60 to 65 ppt. Cintron *et al* (1978) found more dead than living *Rhizophora* trees wherein terrestrial soil salinity exceeded 65 ppt. Teas (1979) reported that transpiration in *Rhizophora* seedling ceases above 65 ppt. On the other hand, members of *Avicennia* who possess salt excretion and limited salt exclusion mechanisms, can exist under more hypersaline conditions. In areas, where there are pronounced draught periods and salinities are high, *Avicennia* may be dominant. *Avicennia marina* shows a wide tolerance from 90 ppt. to near the tidal limit where the salinity is low. The occurrence of *Sonneratia alba* at the seaward edge of the mangrove is due to its preference for normal salinity. The species *Rhizophora stylosa* prefers high salinity whereas *R. apiculata* grows but in less saline areas. *Aegiceras corniculatum* is found in freshwater on the shores of bays along seapage line and at the junction between freshwater and mangrove swamp. So, each species occupies zone of particular salinity.

The xeromorphic features of the mangrove leaves, such as thick cuticle, sunken or chambered stomata, water storage tissue, palisade like mesophyll, bundles terminating in trachieds and mucous cells may serve as a protection against water absorption from highly saline water.

2.2 REPRODUCTIVE STRATEGIES

Nearly all mangroves share two reproductive strategies (Rabinowitz, 1978). These are vivipary (Macnae, 1968; Gill and Tomlinson, 1969) and dispersal by water (van der pijl, 1972). Vivipary means that the embryo develops continuously while attached to parent tree and also during dispersal. Since there is uninterrupted development of the embryo following fertilization to seedling without any intermediate "resting" or true "seed" stage, the word "seed" is inappropriate for viviparous species, such as mangroves, the term "propagule" is generally used in its place.

During the developing time the propagule elongates and the weight distribution changes, so it becomes heavier at the outer, free end. At last, this propagule is dropped from the parent plant and, because of the weight distribution, floats upright in the water. The elongated propagule, however, has the potential to become established in shallow water with tidal influence for its root and to strike the ground. When this happens it puts out roots to anchor it and continues growth into a tree (Fig. 2).

The advantages to this system are obvious for a plant growing on the fringe of the sea. Propagules of *Bruguiera*, *Rhizophora* undergo extensive vivipary while on the tree. When propagules fall from the tree they resemble large green beans. Propagules of all mangroves float and remain viable for extended periods of time.

Members of the *Rhizophora* are usually found at the outer edges of a fringe or bordering tidal creeks and channels. These species have large propagule that can reach some 30 cm long weigh 30-40 gm. Upon settlement the sapling quickly develops supporting roots that protect the young plant from being washed out.

In case of *Bruguiera* and *Rhizophora* this happens in at least two ways: 1) stranding in a vertical position (they float vertically) or 2) stranding in a horizontal position, rooting and then vertical erection by the plant itself. Propagule size and seedling mortality rates are important in considerations of succession and replacement in established mangrove forests (Odum, 1982).

Species of *Avicennia* have small seeds. This is advantageous since the floating propagules can be disseminated by the weak water flows typical of basin forests. The movement of small seeds is not easily hampered by the profuse growth of pneumatophores.

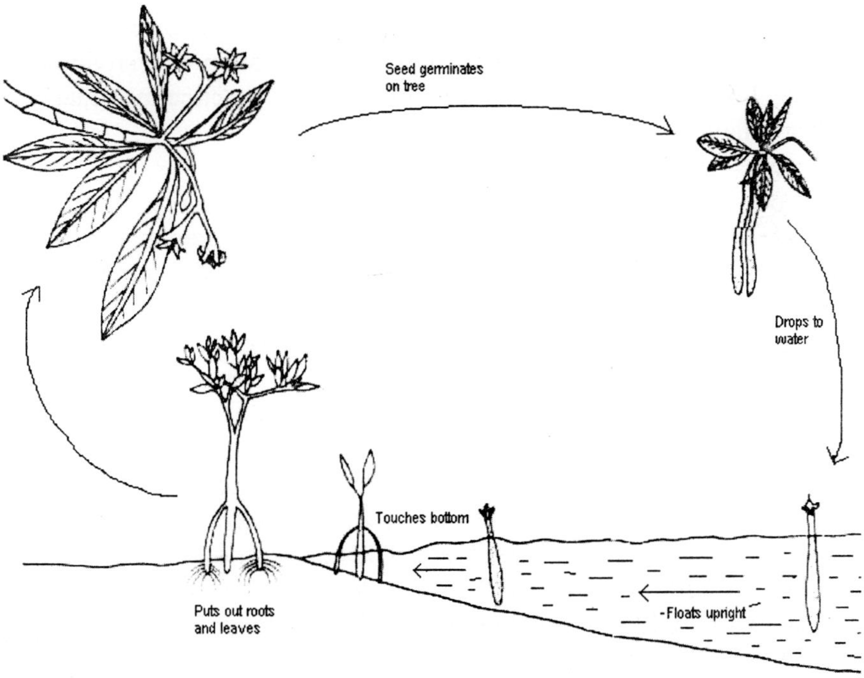

Fig. 2. The life cycle of a typical mangrove tree

Phenology of mangrove species of Andaman and Nicobar Islands has not been studied. In this field, studies on growth and fruiting season are particularly important, as it has many applications such as regeneration, aforestation, plant management, honey analysis, floral biology and the estimation of reproductivity. Author could not get enough information from the local people because they did not observe carefully the phonological phenomena.

2.3. PRIMARY PRODUCTION

Mangrove swamps are one of the richest and most productive areas. Compared to net primary production estimates from other ecosystems, it appears that mangroves are among the World's most productive system. Healthy mangrove ecosystems appear to be more productive than sea grass, marsh grass and most other coastal systems.

Productivity of mangrove estimates come from three methods 1) harvest, 2) gas exchange and 3) litterfall.

Harvest method requires extensive manpower and knowledge of the age of the forest. They are best employed in combination with silviculture practice. Forest Department has recently used this practice in Andaman-Nicobar Islands.

Wood production of mangroves appears to be high compared to other temperate and tropical trees. Balachandran (1988) estimated that the wood production from the north Andaman, middle Andaman and south Andaman mangrove forests averaged 115 m^3/ha, 105 m^3/ha and 59 m^3/ha, respectively. He also pointed out that *Bruguiera* contributes more to be yield in north and middle Andaman. This conclusion assumes that the plants occur within the zone for which they are best adapted.

Mall *et al* (1982) mentioned the density of trees per hectare at Wandoor and Chiriatapu to be 2,507 and 1,167 with stem cover of 0.00195 ha/ha and 0.00964 ha/ha, respectively. The number of anchor roots was 65 per meter with an average circumference of 9.1 cm, resulting in Root Area Index of 0.0428 ha/ha in contrast with stem cover to 0.00076 ha/ha at Wandoor.

Litterfall refers to leaves, wood (twigs), leaf scales, propagules, bracts, and flowers, which fall from the tree. Mangrove leaves are shed continuously throughout the year although a minor peak occurs between the month of August and October. Litterfall is an important ecosystem process because it forms the energy basis for detritus - based food webs in mangrove swamps.

Dagar & Sharma (1989) studied the rate of mangrove litter production at their different sites of Andamans e.g. comparatively less, moderately and highly disturbed area. They found that the litter production amounted to 830 g dry wt./m^2/year, at comparatively less disturbed area, 638 g dry wt./m^2/year at moderate disturbed area and 472 g dry wt./m^2/year at highly disturbed area. They also noted that at one site *Avicennia, Bruguiera, Excoecaria* and *Rhizophora* could contribute 840, 1000, 1100, 950 g dry wt./m^2/year, respectively.

Concerning the primary productivity of phytoplankton, a couple of studies on this aspect had been done in Andaman-Nicobar. Rajagopalan (1987) determined the phytoplankton biomass and the productivity of Andamans water. The result showed that primary productivity ranged from 0.2 to 0.5 g/c/m^2/day at Rangat, Mayabander, Ariel Bay, Henry Lawrence and Neil Islands of north Andaman, while the productivity slightly higher ranged between 0.5-1.0 g/c/m^2/day at Chiraitapu, Wandoor, Corbyn's cove, Kimios Bay of south Andaman and Noncowry of Nicobar Islands. Ramachandran Nair and Gopinathan (1983) also reported the very high production of 2.0 to 3.6 g/c/m^2/day in the mangrove adjacent to mud flats and creeks in Phoenix Bay and Corbyn's Cove of south Andaman.

It is to be noted that the mangrove and adjacent water bodies may be considered very productive in comparison to the offshore waters of Bay of Bengal where the average production is 0.16 g/c/m^2/day (Radhakrishnan, 1978).

Chapter 3

FLORISTICS

3.1 CATEGORIZATION

Taxonomists often face problems in deciding what is and what is not a mangrove, specially when some of the plant species in coastal forests make their way to the mangrove community. Some genera like *Barringtonia, Pemphis*, *Salacia* etc. are in this category.

A total of 58 species of mangrove macrophytes have been recorded from Andaman and Nicobar Islands (Table-3). Among which some e.g. *Aegialitis rotundifolia*, *Aglaia cucullata*, *Rhizophora lamarckii*, *Sonneratia griffithii* and *Xylocarpus mekongensis* could not be located further after the first record. It needs to be confirmed whether these species have become extinct or misidentified. The recorded 58 species included plants that occur on land at the limit of salinity influence i.e. at the transition between the mangrove forest and the back vegetation, fresh water swamp or peat water swamp forest. On the basis of their habitat the Andaman-Nicobar mangrove elements can be classified into two groups.

I. Obligate or exclusive group consists of 36 species in 21 genera, which comprises plants absolutely bound to salty or brackish water (Table-3).

II. Facultative or fringe or back mangrove consists of 22 species in 17 genera, which comprises plants belonging to the littoral vegetation which regularly make their appearances in the back mangroves (Table-3).

Structure

On the basis of the height of plants three categories of forest stratification are observed. The widest trunk with spreading crown is found in the species of *Sonneratia*. The species of *Avicennia* have relatively less spreading crown but in height they are close to *Sonneratia*. The species of *Xylocarpus* represent crown structure between *Sonneratia* and *Avicennia*. These three genera together with *Rhizophora apiculata*,

R. mucronata, *Bruguiera gymnorrhiza* constitute the top canopy of the forest. Shrubs and small trees e.g. *Aegiceras corniculatum, Bruguiera cylindrica, B. parviflora, Excoecaria agallocha*, species of *Ceriops* constitute the second category. The third one is occupied by small shrubs and ferns e.g. species of the genus *Acanthus*, *Acrostichum aureum, Aegialitis rotundifolia, Nypa fruticans.*

Table-3: List of Mangrove species in Andaman-Nicobar Islands

I. OBLIGATE OR EXCLUSIVE GROUP
(Plants are absolutely bound to salty or brackish water)

Family	Scientific Name
Adiantaceae	*Acrostichum aureum* Linn. *A. speciosum* Willd.
Arecaceae (Palmae)	*Nypa fruticans* Wurmb. *Phoenix paludosa* Roxb.
Asclepiadaceae	*Sarcolobus globosus* Wall.
Avicenniaceae	*Avicennia marina* (Forsk.) Vierh. *A. officinalis* Linn.
Bignoniaceae	*Dolichandrone spathacea* (L.*f.*) K. Schum.
Combretaceae	*Lumnitzera littorea* (Jack.) Voigt *L. racemosa* Willd.
Euphorbiaceae	*Excoecaria agallocha* Linn.
Fabaceae	*Derris trifoliata* Lour.
Meliaceae	*Aglaia cucullata* (Roxb.) Pellegrin *Xylocarpus gangeticus* (Prain) C.E. Parkinson *X. granatum* Koen. *X. mekongensis* Pierre *X. moluccensis* (Lamk.) Roem.
Myrsinaceae	*Aegiceras corniculatum* (Linn.) Blanco
Plumbaginaceae	*Aegialitis rotundifolia* Roxb.
Rhizophoraceae	*Bruguiera cylindrica* (Linn.) Bl. *B. gymnorrhiza* (Linn.) Savigny *B. parviflora* (Roxb.) Wt. & Arn. *ex* Griff. *B. sexangula* (Lour.) Poir. *Ceriops decandra* (Griff.) Ding Hou *C. tagal* (Perr.) C.B. Rob. *Kandelia candel* (Linn.) Druce *Rhizophora apiculata* Blume

Family	Scientific Name
	R. lamarckii Montr. *R. mucronata* Lamk. *R. stylosa* Griff.
Rubiaceae	*Scyphiphora hydrophyllacea* Gaertn. *f.*
Sonneratiaceae	*Sonneratia alba* J. Smith *S. caseolaris* (Linn.) Engl. *S. griffithii* Kurz
Sterculiaceae	*Heritiera littoralis* Dryand
Tiliaceae	*Brownlowia tersa* (Linn.) Kosterm.

II. FACULTATIVE OR FRINGE OR BACK MANGROVE
(Plants belonging to the littoral vegetation regularly make their appearances in the back mangroves)

Family	Scientific Name
Acanthaceae	*Acanthus ebracteatus* Vahl *A. illicifolius* Linn. *A. volubilis* Wall.
Apocynaceae	*Cerbera manghas* Linn. *C. odollum* Gaertn.
Asclepiadaceae	*Finlaysonia obovata* Wall. *Tylophora tenuissima* (Roxb.) W. & A. *ex* W.
Asteraceae	*Pluchea indica* (Linn.) Less.
Caesalpiniaceae	*Caesalpinia bonduc* (Linn.) Roxb. *C. crista* Linn.
Cyperaceae	*Fimbristylis ferruginea* (Linn.) Vahl
Fabaceae	*Cynometra iripa* Kosteletzky *C. ramiflora* Linn. *Dalbergia candenatensis* (Dennst.) Prain
Flagellariaceae	*Flagellaria indica* Linn.
Hippocrateaceae	*Salacia chinensis* Linn.
Lecythidaceae	*Barringtonia racemosa* (Linn.) Spreng.
Lythraceae	*Pemphis acidula* J.R. & G. Forst.
Malvaceae	*Hibiscus tiliaceous* Linn.
Myrsinaceae	*Ardisia solanacea* Roxb.
Pandanaceae	*Pandanus odoratissimus* Linn. *f.*
Verbenaceae	*Clerodendrum inerme* (Linn.) Gaertn.

Taxonomic Note

On *Cerbera manghas* and *C. odollum*

These two occasional species of coastal forest and back mangal communities are very closely related and have been much confused. As a result different authors have used the names *manghas* and *odollum* for both entities, despite the clear distinction between them. So there are controversies regarding the delimitation the proper status of these two species.

After careful observation of the specimens of *Cerbera manghas* and *C. odollum* of Andaman-Nicobar Islands, it was found that though the two species are vegetatively somewhat close to each other, yet, a few morphological characters show the differentiation between them. These two species may be distinguished as follows:

C. odollum Gaertn.	*C. manghas* Linn.
Leaves apiculate with a fine point.	Leaves bluntly pointed.
Primary veins perpendicular to midrib.	Primary veins arcuate.
Corolla with a yellow eye, throat of the corolla not closed by hairy connective ridges.	Corolla with a red eye, throat of the corolla is closed by densely hairy connective ridges.
Corolla tube 1.5 - 2.0 cm long.	Corolla tube 1.5 - 4.0 cm long.
Stamens inserted at or shortly above middle of locally widened tube.	Stamens inserted at the mouth of tube.
Fruits with endocarp somewhat transitional to fibres of inner mesocarp.	Fruits without clear distinction between mesocarp and endocarp.

On *Rhizophora lamarckii* Montr.

Singh *et al* (1987) recorded this species in their publication "A new record of some mangrove species from Andaman Islands and their distribution" from Havelock Islands (South Andaman) where they have reported to have seen only some trees of this species amidst *R. apiculata* and *R. mucronata*. They have provided brief description with a sketch of the plant, which indicates this to be *R. lamarckii*. Unfortunately no voucher specimens are indicated. Lakshmanan & Rajeshwari (1983) first recorded this taxon for India from Pichavaram of Tamil Nadu.

R. lamarckii was first described from Canala, the north west coast of New Caledonia and was considered for long to be endemic to this area. Later its distribution in Queensland, Papua New Guinea, the Solomon Islands and the New Hebrides i.e. between longitude 150° E and 180° E (Tomilson, 1978 & 1986) is known.

Based on the circumstantial evidences, such as poor representation in the area of occurrence amidst well recongnized species *R. apiculata* and *R. stylosa*, extremely rare or lack of occurrence of seedlings, poor seed formation and aberrations in stamen morphology, all indicating sterility, it is concluded that *R. lamarckii* is a hybrid species (Tomilson, 1978; Muniyandi & Natarajan, 1985; Lakshmanan & Poornima, 1988). Since this species is occurring amidst *R. apiculata* & *R. stylosa* in the type locality and elsewhere, it is reported to be a possible sterile F_1 hydrid between them (Tomilson, 1986).

The rare collection of a seedling of *R. lamarckii* although once only by Muniyandi & Natarajan (1985) and description of seedlings (with emerging hypocotyl) by Singh *et al* (1987) indicate that sterility is not absolute. Further observations of viable seedlings and cytological studies may throw more light on genetic status of the species.

Identification

The following at a glance key is intended as a preliminary aid to the identification of mangrove species of Andaman & Nicobar Islands.

Key to the species

1a. Flowering plants 2

1b. Non-flowering plants 57

2a. Leaves with reticulate venation; flowers usually 4-5 merous; embryo with two cotyledons 3

2b. Leaves with parallel venation; flowers usually 3-merous; embryo with one cotyledon 53

3a. Latex present 4

3b. Latex absent... 9

4a. Leaves opposite 5

4b. Leaves alternate 7

5a. Corona absent; seeds without coma.................................... *Sarcolobus globosus*

5b. Corona present; seeds with coma 6

6a. Leaves with trichomes; calyx on the inside with 5 basal glands; filaments connate in a short tube; top of stigma with a central knob........................ *Tylophora tenuissima*

6b. Leaves without trichomes; calyx without glands; filaments free; stigma flat topped .... *Finlaysonia obovata*

7a. Leaves 3 - 8 cm long, margin obscurely notched, 2 basal glands usually 1 on each side of blade at its insertion on the petiole; inflorescences catkin like spikes; flowers 2.0 - 3.5 cm long; fruits 3-lobed capsule, 7 mm in diameter .... *Excoecaria agallocha*

7b. Leaves 10 - 30 cm long, margin entire, no basal glands present on the blade; inflorescences terminal paniculate cymes; flowers 4 - 7 cm long; fruits drupaceous with fibrous pericarp, 3 - 4 cm in diameter.. 8

8a. Leaves apiculate, with a fine point, primary veins perpendicular to midrib; corolla with a yellow eye, tube short (1.5 - 2.0 cm), free corolla segments long (2.0 - 3.5 cm), throat of the corolla not closed by sparsely hairy connective ridges .... *Cerbera odollum*

8b. Leaves bluntly pointed, primary veins arcuate; corolla with a red eye, tube long (1.5 - 4.0 cm), free segments short (2.0 - 2.5 cm); throat of the corolla closed by densely hairy connective ridges.. *Cerbera manghas*

9a. Leaves simple... 10

9b. Leaves compound 42

10a. Leaves opposite 11

10b. Leaves alternate... 33

11a. Stipules present 12

11b. Stipules absent 23

12a. Stipules interpetiolar, not overwrapping, persistent .... *Scyphiphora hydrophyllacea*

12b. Stipules overwrapping, standing above the leaf insertion, falling to leave an annual nodal scar 13

13a. Trees with extensively developed stilt roots; flowers tetramerous; calyx 4-lobed; petals 4, entire, without distal appendages; anthers sessile, multilocellate, dehiscing by an adaxial valve 14

13b. Trees without stilt roots; flowers penta or more (upto 16) merous; calyx 5 or more lobed; petals 5 or more, deeply emarginate, with apical appendages; anthers with filaments, 4-locular, dehiscing by longitudinal ruptures 17

14a. Flowers in 2's; peduncle *ca* 10 mm long, shorter than petiole, each flower sessile with a pair of massive cupular bracteoles; petals glabrous; stamens 12, sessile ... . *Rhizophora apiculata*

14b. Flowers in 4's or more; peduncle 15 - 40 mm long, longer than petiole; each flower with a pedicel; bracteoles green, fleshy but never corky; petals slightly to conspicuously hairy on the margins; stamens 8 or about 16, with short filaments 15

15a. Flowers usually in 4's; peduncle about 15 mm long, stout; petals with inconspicuous marginal hairs; stamens variable in number (8-15-) 16; styles 2 - 3 mm long .... *Rhizophora lamarckii*

15b. Flowers 4, 8 or more per inflorescence; peduncle 25 - 40 mm long, slender; petals with conspicuous marginal hairs; stamens always 8; styles either long or absent 16

16a. Stigma sessile; free part of ovary emerging beyond the disc; seedling warted .... *Rhizophora mucronata*

16b. Stigma on a slender style 4 - 5(-6) mm long; free part of ovary enclosed by the disc; seedling smooth.. *Rhizophora stylosa*

17a. Trees without pneumatophores; petals without marginal hairs; stamens more than twice as many as calyx lobes; hypocotyl of seedling slender, gradually narrowed and distantly pointed .... *Kandelia candel*

17b. Trees with knee like pneumatophores; petals with marginal hairs; stamens twice as many as calyx lobes; hypocotyl of seedling blunt or abruptly pointed distally ... 18

18a. Leaves usually less than 10 cm long, apex rounded; calyx 5-or 6-lobed, blunt, 2 - 4 mm long; petals less than 0.5 cm long; stamens 10 or 12 19

18b. Leaves usually more than 10 cm long, apex acute; calyx 8-16 lobed, lobes pointed, upto 15 mm long; petals 0.5 - 2 cm long; stamens 16 - 32............................ 20

19a. The apex of petal with 3 clavate appendages; stamens with long, slender filaments, much exceeding the blunt anther; hypocotyl long, upto 25 cm or more, warty throughout, terete or slightly ridged .... *Ceriops tagal*

19b. The apex of petal with a fringe of filamentous appendages; stamens with a short filament, equal length to anther; hypocotyl not more than 15 cm long, warty towards apex, sharply ridged.. *Ceriops decandra*

20a. Flowers (2-) 3 (-5) in pedunculate cymes, less than 2 cm long 21

20b. Flowers solitary, 3-4 cm long 22

21a. Calyx tube smooth; lobes stout, long (more than 3 mm), half the length of the calyx, horizontal or reflexed in fruit .... *Bruguiera cylindrica*

21b. Calyx tube ribbed; lobes slender, short (less than 3 mm), only one-fourth to one-fifth the length of the calyx, erect or at most slightly spreading in fruit... *Bruguiera parviflora*

22a. Petal lobes blunt, without filamentous appendages .... *Bruguiera sexangula*

22b. Petal lobes acute, each extended into 3 filamentous appendages.. *Bruguiera gymnorrhiza*

23a. Petals more or less united 24

23b. Petals free from each other 29

24a. Plants with leaves and nodes spinescent; ovule more than one in each locule 25

24b. Plants not spinescent; ovule solitary in each locule 27

25a. Bracteoles persistent, upto 1 cm long in fruit; ripe fruits 2.5 - 3.0 cm long; seeds *ca* 10 mm in diameter... *Acanthus illicifolius*

25b. Bracteoles inconspicuous, small, early deciduous or absent; ripe fruits less than 2.0 cm long; seeds 5-7 mm in diameter... 26

26a. Bracts shorter than the calyx; bracteoles present but early deciduous... *Acanthus ebracteatus*

26b. Bracts longer than the calyx; bracteoles never present................ *Acanthus volubilis*

27a. Shrubs (upto 3 m tall) without pneumatophores; corolla with a long tube (24 - 30 mm); stamens exserted .... *Clerodendrum inerme*

27b. Trees (8-25 m tall) with pneumatophores; corolla with very short tube (1-2 mm long); stamens very shortly exserted during full anthesis 28

28a. Leaf blades ovate or lanceolate to lanceolate-oblong or elliptic, usually abruptly acute at the apex; expanded corolla 5-8 mm wide............................ *Avicennia marina*

28b. Leaf blades obovate or broadly oblong, rounded at the apex; expanded corolla 12-15 mm wide.. *Avicennia officinalis*

29a. Stamens more than twice as many as petals... 30

29b. Stamens not more than twice as many as petals.. 32

30a. Calyx cup shaped, enclosing the base of the fruit, prominently ribbed.....................
.... *Sonneratia alba*

30b. Calyx flat, extended horizontally, not enclosing the fruit, at most obscurely ribbed
.... 31

31a. Twigs not pendulous; leaves obovate to suborbicular; veins conspicuous, prominent on the upper surface; petals absent; filaments white................ *Sonneratia griffithii*

31b. Twigs pendulous, slender; leaves elliptic-oblong or oval-obovate to narrowly elliptic; veins inconspicuous; petals usually present; filaments red below, white distally
.... *Sonneratia caseolaris*

32a. Woody climbers; leaves 5-9 cm long; sepals free; petals 4-5; stamens 3; filaments flattened.. *Salacia chinensis*

32b. Herbs; leaves 1.0 - 3.5 cm long; sepals united into a tube; petals 6; stamens 12; filaments not flattened... *Pemphis acidula*

33a. Plants with either stilt roots or with large buttresses 34

33b. Plants without stilt roots and buttresses 37

34a. Plants with stilt roots; buttresses absent 35

34b. Plants without stilt roots; buttresses present .... *Heritiera littoralis*

35a. Anthers with transverse septa; style 1 36

35b. Anthers without transverse septa; styles 5 .... *Aegialitis rotundifolia*

36a. Leaves with pellucid or coloured dots; corolla reddish violet .... *Ardisia solanacea*

36b. Leaves without any dots; corolla white .... *Aegiceras corniculatum*

37a. Leaves stipulate 38

37b. Leaves exstipulate 40

38a. Shrubs, 2-3 m tall; leaves dark-green and glossy above, densely dull silvery scaly beneath ... . *Brownlowia tersa*

38b. Trees, 5-15 m tall; leaves not silvery scaly beneath 39

39a. Leaves crenulate; flowers in pendulous raceme; petals 4, pink to red; fruits with persistent calyx... *Barringtonia racemosa*

39b. Leaves entire; flowers axillary or in terminal erect raceme; petals 5, yellow with purple base; fruits without persistent calyx .... *Hibiscus tiliaceous*

40a. Flowers in heads, heads corymbose, provided with a basal involucre... *Pluchea indica*

40b. Flowers in erect racemes, not in head, not provided with involucre........................41

41a. Plants with knee shaped pneumatophores; inflorescences terminal; flowers pedicellate; petals red; stamens twice as long as the petals..................... *Lumnitzera littorea*

41b. Plants without pneumatophores; inflorescences axillary; flowers sessile; petals white; stamens approximately equalling the petals....................... *Lumnitzera racemosa*

42a. Leaves imparipinnate 43

42b. Leaves paripinnate 46

43a. Leaflets opposite 44

43b. Leaflets alternate 45

44a. Trees upto 20 m tall; flowers regular; fruit not a legume or lomentum..................... .... *Dolichandrone spathacea*

44b. Woody climbing shrubs; flowers irregular; fruit a legume or lomentum .... *Derris trifoliata*

45a. Trees; leaves 30-40 cm long; leaflets 9-19 x 4-6 cm, terminal leaflet forming a cup at base; flowers yellow; fruits globose.. *Aglaia cucullata*

45b. Climbers; leaves 4.0 - 12.5 cm long; leaflets 1.7-5.0 x 1.0-2.6 cm; terminal leaflet not as above; flowers white; pods flat .... *Dalbergia candenatensis*

46a. Leaflets opposite or subopposite 47

46b. Leaflets alternate 50

47a. Climbers with numerous recurved-prickles 48

47b. Trees without recurved prickles... 49

48a. Stipules conspicuous; leaflets opposite or subopposite, 16-24 per pinna; flowers unisexual; fruits armed with rigid spines; seeds grey .... *Caesalpinia bonduc*

48b. Stipules obscure; leaflets always opposite, 2-4 pairs per pinna; flowers bisexual; fruits unarmed; seeds black .... *Caesalpinia crista*

49a. Sepals curved distally when reflexed at anthesis; style bent, oblique to the dorsal line of the ovary, becoming a prominent lateral beak in fruit; ovary wall hairy inside .... *Cynometra iripa*

49b. Sepals not curved distally when reflexed at anthesis; style straight, in line with the dorsal part of the ovary; fruits without a prominent lateral beak; ovary wall glabrous inside .... *Cynometra ramiflora*

50a. Buttresses present; staminal teeth prominent 51

50b. Buttresses absent; staminal teeth obscure 52

51a. Stem smooth; leaflets obovate; fruits 20-35 cm in diameter.. *Xylocarpus granatum*

51b. Stem rough and fissured; leaflets elliptic-oblong; fruits 7-13 cm in diameter.. *Xylocarpus gangeticus*

52a. Leaflets ovate; stigma cupular; fruits 5-7 cm in diameter..... *Xylocarpus moluccensis*

52b. Leaflets elliptic-oblong; stigma descoid; fruits 8-15 cm in diameter.. *Xylocarpus mekongensis*

53a. Perianth absent or represented by hypogynous setae, scales or lodicules; flowers reduced to spikelets .... *Fimbristylis ferruginea*

53b. Perianth present; flowers not reduced to spikelets 54

54a. Climbers; leaves ending in a short coiled tendril .... *Flagellaria indica*

54b. Small trees or shrubs; leaves not ending in a tendril 55

55a. Di- or trichotomously branched large shrubs; leaves trifariously spiralled, gathered in terminal crowns .... *Pandanus odoratissimus*

55b. Unbranched small trees or shrubs; leaves pinnate, with numerous linear sword shaped leaflets 56

56a. Stemless palm; leaf segments lanceolate, plicate; fruits hexagonal....... *Nypa fruticans*

56b. Sobiliferous palm, stem 1-4 m high; leaf segments ensiform with induplicate margin; fruits oblong, terete .... *Phoenix paludosa*

57a. Apex of sterile pinnae abruptly rounded .... *Acrostichum aureum*

57b. Apex of sterile pinnae narrowly acuminate............................. *Acrostichum speciosum*

Key to the families

1a. Flowering plants 2

1b. Non-flowering plants .... 2. Adiantaceae

2a. Flowers 4-5 merous; seeds with two cotyledons. 3

2b. Flowers usually 3-merous; seeds with one cotyledon.. 25

3a. Latex present 4

3b. Latex absent 6

4a. Leaves opposite .... 5. Asclepiadaceae

4b. Leaves alternate 5

5a. Leaves 10-30 cm long; inflorescences terminal paniculate cymes; fruits drupaceous with fibrous pericarp, 3-4 cm in diameter.. 3. Apocynaceae

5b. Leaves 3-8 cm long; inflorescences catkin like spikes; fruits 3-lobed capsules, 7 mm in diameter. ....12. Euphorbiaceae

6a. Leaves simple 7

6b. Leaves compound 23

7a. Leaves opposite 8

7b. Leaves alternate 15

8a. Stipules present 9

8b. Stipules absent ... .10

9a. Stipules interpetiolar, persistent .... 24. Rubiaceae

9b. Stipules overwrapping, standing above the leaf insertion, falling to leave an annual nodal scar .... 23. Rhizophoraceae

10a. Petals more or less united ... .11

10b. Petals free from each other 13

11a. Plants with leaves and nodes spinescent; ovules more than one in each locule .... 1. Acanthaceae

11b. Plants not spinescent; ovule solitary in each locule... 12

12a. Trees, 8-25 m tall, with pneumatophores ....7. Avicenniaceae

12b. Shrubs, upto 3 m tall, without pneumatophores .... 28. Verbenaceae

13a. Stamens more than twice as many as petals .... 25. Sonneratiaceae

13b. Stamens not more than twice as many as petals 14

14a. Herbs; sepals united into a tube; petals 6; stamens 12 .... 17. Lythraceae

14b. Woody climbers; sepals free; petals 4-5; stamens 3 ....15. Hippocrateaceae

15a. Plants with either stilt roots or with large buttresses 16

15b. Plants without stilt roots or buttresses 19

16a. Plants with stilt roots, buttresses absent 17

16b. Plants without stilt roots, buttresses present 18

17a. Anthers with transverse septa; style one .... 20. Myrsinaceae

17b. Anthers without transverse septa; styles five .... 22. Plumbaginaceae

18a. Leaves on the undersurface covered with golden fimbriate scales; filaments united into a column with 5 fertile stamens alternating with staminodes; stigma simple .... 26. Sterculiaceae

18b. Leaves glabrous and without scales; filaments united into a tube, stamens not as above; stigma capitate or discoid.. 19. Meliaceae

19a. Leaves stipulate 20

19b. Leaves exstipulate 22

20a. Shrubs, 2-3 m tall; leaves densely dull silvery scaly beneath .... 27. Tiliaceae

20b. Trees, 5-15 m tall; leaves not silvery scaly beneath 21

21a. Leaves entire; flowers axillary or in terminal erect raceme; petals 5; fruits without persistent calyx.. 18. Malvaceae

21b. Leaves crenulate; flowers in pendulous raceme; petals 4; fruits with persistent calyx .... 16. Lecythidaceae

22a. Flowers in heads, heads corymbose, provided with involucre .... 6. Asteraceae

22b. Flowers in erect racemes, not in head, not provided with involucre .... 10. Combretaceae

23a. Leaves imparipinnate 24

23b. Leaves paripinnate .... 9. Caesalpiniaceae

24a. Woody climbing shrubs or trees; flowers irregular; fruit a legume or lomentum .... 13. Fabaceae

24b. Trees; flowers regular; fruit not a legume or lomentum.................... 8. Bignoniaceae

25a. Perianth absent or represented by hypogynous setae, scales or lodicules; flowers reduced to spikelets.. 11. Cyperaceae

25b. Perianth present; flowers not reduced to spikelets.. 26

26a. Climbers; leaves ending in a short coiled tendril .... 14. Flagellariaceae

26b. Small trees or shrubs; leaves not ending in a tendril 27

27a. Di-trichotomously branched large shrubs; leaves trifariously spiralled, gathered in terminal crowns .... 21. Pandanaceae

27b. Unbranched small trees or shrubs; leaves pinnate with numerous, linear, sword shaped leaflets .... 4. Arecaceae

1. ACANTHACEAE

Acanthus Linn.

Key to the species

1a. Bracteoles persistent, upto 1 cm long; ripe fruits 2.5 - 3.0 cm long........................... ....**2. *A. illicifolius***

1b. Bracteoles inconspicuous, early deciduous or absent; ripe fruits less than 2 cm long **2**

2a. Bracts shorter than the calyx; bracteoles present but early deciduous........................ ....**1. *A. ebracteatus***

2b. Bracts longer than the calyx; bracteoles never present...............................**3. *A. volubilis***

1. Acanthus ebracteatus Vahl, Symb. Bot. 2:75, t. 40. 1791; Parkinson, For. Fl. Andam. 217. 1923.

Erect undershrubs. Leaves opposite, oblong or elliptic, with or without marginal spines, always with an apical spine, glabrous, coriaceous. Spikes terminal, many flowered, lax to dense; bracts minutely ciliate, glabrous on dorsal side; bracteoles subtending the calyx, absent or fugacious; calyx deeply 4-partite, anterior and posterior segments largest; corolla tube short, on the inside top with a ring of hairs, 3-lobed, lower lip large, ovate-obovate, white; stamens 4, in the throat; anthers medifixed; ovules 2 in each ovary cell, superposed; style bidentate. Capsules ovoid-oblong, compressed, coriaceous. Seeds tuberculate-rugose, glabrous.

Fl. March - June; *Fr.* June - August.

Distr. India : Andaman Islands.

Indo-China, Malay Peninsula and Archepelago to New Guinea.

Notes. Common along tidal streams, inland borders of mangrove swamps under the influence of salt or brackish water.

2. Acanthus illicifolius Linn., Sp. Pl. 639. 1753; Parkinson, For. Fl. Andam. 217. 1923.

Erect, ascending or rarely scandent shrubs up to 2 m tall. Stems with stilt roots. Leaves oblong or lanceolate, pinnatifid or toothed, always with an apical spine, narrowed at base, often with marginal spines, glabrous, shining. Flowers in terminal or pseudo-axillary dense strobilate spike, sessile; bracts ovate, acute, entire, glabrous; bracteoles 2, subtending the calyx, persistent, calyx 4-segmented, segments glabrous, shortly connate in two opposite pairs; corolla 5-segmented, segments connate, 2-lipped, hairy outside, blue or bluish violet, on the inside top with a ring of hairs; stamens with thick filaments; anthers densely bearded. Capsules ovoid-oblong, compressed, apiculate, shining. Seeds kidney shaped.

Fl. April - June; *Fr.* June - August.

Distr. India : Andaman & Nicobar Islands, Sunderbans, East & West Coast mangrove areas.

Tropical Asia, S.E. Africa through Malaysia to Polynesia.

Notes. Gregarious in brackish swamps along the seashore and tidal streams.

3. **Acanthus volubilis** Wall., Pl. As. Rar. 2. 56. t. 172. 1831; Parkinson, For. Fl. Andam. 217. 1923.

Twining shrubs, up to 8 m tall. Leaves oblong or obovate-oblong, entire, obtuse or mucronulate at apex, cuneate at base. Flowers in simple or branched spikes; bracts lanceolate, subtending the calyx, caducous; calyx 4-lobed, shortly cuneate below in two opposite pairs, outer lobes larger than the inner; corolla 5-lobed, lobes connate, 2-lipped, white, lower lip shortly 3-lobed; stamens 4, didynamous; anthers densely bearded. Capsules ellipsoid, compressed, mucronate at apex.

Fl. April - June; *Fr.* June - August.

Distr. India : Andaman & Nicobar Islands, Sunderbans.

Burma, Thailand, Java, Malaysia & Australia.

Notes. Rare in tidal forests and swamps, reported to grow in sheltered mangrove areas.

2. ADIANTACEAE

Acrostichum Linn.

Key to the species

1a. Apex of sterile pinnae abruptly rounded; juvenile leaves with an oblong, blunt blade .... **1. *A. aureum***

1b. Apex of sterile pinnae narrowly acuminate; juvenile leaves with a lanceolate, pointed blade .. .. **2. *A. speciosum***

1. Acrostichum aureum Linn., Sp. Pl. 2:1069. 73. 1753.

Stock stout, erect. Scales on rhizome. Stipe and frond together 3 to 4 m tall; pinnae many, simple, only the upper ones fertile. Sterile pinnae 50 x 10 cm, on 2 cm long stalks, base cuneate to rounded, apex abruptly rounded to truncate and shortly mucronate. Fertile pinnae like the sterile but smaller.

Distr. This is a pantropical species, which grows always in mangrove swamps.

Notes. This species is extremely abundant in mangrove areas in all parts of Andaman & Nicobar Islands and can withstand partial immersion by the tide each day. It grows sometimes near the sea. It is found in the landward side of mangrove, surviving sometimes in ditches or in wet ground, which have been completely cut off from the sea.

2. Acrostichum speciosum Willd., Sp. Pl. 5:117. 1810.

Stipe and frond together to about 1.5 m tall. Sterile pinnae 30 x 4 cm, on 1 cm long stalks, base cuneate, apex narrowly acuminate.

Distr. Confined to tropical Asia and Australia.

Notes. This species is abundant in mangrove areas throughout Andaman & Nicobar Islands, occurring in the parts more frequently inundated by the tide, growing specially on heaps of mud raised by prawns above the general level of the ground.

3. APOCYNACEAE

Cerbera Linn.

Key to the species

1a. Leaves apiculate with a fine point, primary veins perpendicular to midrib; corolla with a yellow eye, tube short, but free corolla segments long, throat of the corolla not closed by sparsely hairy connective ridges .... **2. *C. odollum***

1b. Leaves bluntly pointed, primary veins arcuate; corolla with a red eye, tube long, but free corolla segments short, throat of the corolla closed by densely hairy connective ridges .... **1. *C. manghas***

1. Cerbera manghas Linn., Sp. Pl. 208. 1753.

Trees, 4-6 m high, with milky latex. Leaves alternate, closely set or whorled at the apices of branchlets, ovate-oblong or oblanceolate, acuminate at apex, base rounded. Flowers in terminal paniculate cymes, large, white, funnel shaped, 5-merous; calyx deeply divided; corolla salver shaped, with a red eye, tube widened only at the top, tops of antheriferous ridges ending very near the throat, closing the entrance to the tube, densely long woolly. Fruits globose-ovoid-ellipsoid, drupaceous, with fibrous pericarp.

Fl. March - June; *Fr.* June - August.

Distr. India : Andaman & Nicobar Islands, East & West Coast mangrove areas.

Sri Lanka, Malaysia, China, Australia.

Notes. Common along the intertidal banks of creeks and channels in mangrove forests.

2. Cerbera odollum Gaertn., Fruct. 2:193, t. 124. fig. 1. 1791; Parkinson, For. Fl. Andam. 206. 1923.

Trees, 7-15 m high, with milky juice. Leaves alternate, obovate-lanceolate to oblong-lanceolate, shortly acuminate at apex, tapering acute at base. Flowers in terminal paniculate cymes, white, 5-merous; calyx segments recurved; corolla with a yellow eye, tube widened upwards from just above the middle, tops of antheriferous ridges ending far below the throat, not closing the entrance of the tube, not densely hairy. Fruits globose-ovoid or ellipsoid, drupaceous.

Fl. Jan. - July; *Fr.* Oct. - Jan.

Distr. India : Andaman & Nicobar Islands.

Vietnam, Java.

Notes. Common in intertidal banks.

4. ARECACEAE

Key to the genera

1a. Stemless palms; leaf segments lanceolate, plicate; fruits hexagonal .... **1. *Nypa***

1b. Sobiliferous palms, stem 1-4 m high; leaf segments ensiform with induplicate margin; fruits oblong, terete .... **2. *Phoenix***

1. Nypa Steck

Nypa fruticans Wurmb. in Verh. Batav. Geneotsch. Kunsten. 1:350. 1781; Parkinson, For. Fl. Andam. 263. 1923.

Stemless palms, rootstock prostrate. Leaves arising from the rootstock, pinnatisect; leaflets linear-lanceolate. Spathes many; spadix terminal, erect in flower, drooping in fruit. Male flowers minute, brown in colour, arranged in catkins on lateral branches of the spadix; female flowers larger, arranged in globose terminal heads, yellow in color. Fruits ovoid or globose, syncarpous, compressed, turbinate, fibrous, with hexagonal, 1-seeded carpels.

Fl. Feb. - June; *Fr.* June - Sept.

Distr. India : Andaman-Nicobar Islands, Sunderbans and East Coast Mangrove areas.

Sri Lanka, Bangladesh, Burma, Malaysia, Philippines to Australia.

Notes. This species is gregariously growing along the sheltered intertidal regions of creeks in the mångrove forests. It prefers low percentage of salt.

2. Phoenix Linn.

Phoenix paludosa Roxb., Fl. Ind. 3:789. 1832; Parkinson, For. Fl. Andam. 263. 1923.

Subarboreous, gregarious, sobiliferous, annulate palms. Stems 1-4 m high, base supported by spongy, needle like, perforated breathing roots. Leaves imparipinnate with many ensiform segments; segments bifarious, waxy, glabrous, acuminate at apex, conduplicate at base. Inflorescences spadix, spadices arising in between leaves. Flowers dioecious, yellowish-white; petals 3; stamens 3 (in male flowers); carpels 3 (in female flowers). Fruits drupaceous, oblong or ellipsoid, turning red to purple when ripe.

Fl. Jan. - May; *Fr.* May - August.

Distr. India : Andaman & Nicobar Islands, Sunderbans, Mahanadi Delta.

Thailand and Vietnam.

Notes. This species is extremely abundant in mangrove areas in all parts of Andaman Islands, but not so abundant in Nicobar Islands. It grows on elevated muddy swamps along estuarine banks and thrives very well near the sea coast. Seems to tolerate and even to thrive better in places with a higher percentage of salt in the water, compared to *Nypa fruticans*.

5. ASCLEPIADACEAE

Key to the genera

1a. Corona absent; seeds without coma .... **2. *Sarcolobus***

1b. Corona present; seeds with coma **2**

2a. Leaves with trichomes; calyx on the inside with 5 basal glands; filaments connate in a short tube; top of stigma with a central knob .... **3. *Tylophora***

2b. Leaves without trichomes; calyx without glands; filaments free; stigma flat topped .... **1. *Finlaysonia***

1. **Finlaysonia** Wall.

Finlaysonia obovata Wall., Pl. As. Rar. 2:48, t. 162. 1831.

Twining shrubs with milky juice. Leaves opposite, narrowly lanceolate to obovate, rounded or emarginate at apex, cuneate at base, glabrous, fleshy. Flowers in peduncled cymes, solitary, foetid; calyx 5-segmented, overlapping to the right, greenish yellow or brownish yellow, with long white hairs inside; corona scales white, inserted on the corolla, with uncinate tops; stamens inserted on the throat; filaments free; anthers with an incurved, broad, acute, apical membrane; ovaries free; stigma quinquangular, flat-topped. Follicles ovoid, narrowed at base, with a hooked acuminate top, ribbed. Seeds ovoid, flattened, with a tuft of silky hairs.

Fl. Oct. - Jan.; *Fr.* Jan. - March.

Distr. India: Andaman & Nicobar Islands, East and West Coast Mangrove areas.

Malaysia to Moluccas.

Notes. Frequent along the seashores in swampy places, often in the mangrove along tidal streams.

2. **Sarcolobus** R.Br

Sarcolobus globosus Wall., Res. 12:568. t. 4. 1818.

Twining lianas, 2-5 m long, glabrescent. Leaves opposite; blade glossy green above, pale below, ovate or elliptic, rarely obovate, apex acute to obtuse, base obtuse to shallowly cordate. Inflorescences pedunculate spirally elongating raceme. Flowers small, starry, crowded; corolla subrotate, tube glabrous, throat and lobes sparsely to densely pubescent, inside with white hairs; gynostegium with the anther wings curving half way down the stamen tube; corolline and staminal corona not present. Fruits globose, base unequal. Seeds many, flattened.

Fl. June - August; *Fr.* August - Sept.

Distr. India : Nicobar Islands, Sunderbans and Mahanadi Delta.

Burma, Thailand, Indo-China, Malaysia, Java and S.E. Borneo.

Notes. It is reported to grow on muddy intertidal areas; often in association with and climbing on *Phoenix paludosa.*

3. **Tylophora** R. Br.

Tylophora tenuissima (Roxb.) W. & A. *ex* W., Contr. Bot. Ind. 49. 1834. *Asclepias tenuissima* Roxb., Fl. Ind. 2:41. 1824. *Tylophora tenuis* Bl., Bijdr. 1062. 1826.

Climbers. Leaves ovate-elliptic-oblong-lanceolate, coriaceous, minutely cuspidate, acute, obtuse, rounded or sub-cordate at base, glabrous or pubescent on the midrib, with basal trichomes. Flowers in repeatedly forked cymes; calyx on the inside with 5 basal glands, 5-parted, segments oval-oblong, pale; corolla 5-parted, segments oval-oblong, dark purple; corona-scales at the outer angle very small, knob-shaped; anther short, with apical membrane over-reaching the stigma; stigma semi-globose, top of stigma with a central knob. Follicles finally more or less in a line with each other, ovoid-lanceolate, acuminate, smooth. Seeds comose.

Fl. April - July; *Fr.* July - Oct.

Distr. India : Andaman & Nicobar Islands, Sunderbans, Mahanadi Delta.

Java and Borneo.

Notes. Common in the back mangrove region associated with *Caesalpinia crista* and *Acrostichum* sp.

6. ASTERACEAE

Pluchea Cass.

Pluchea indica (Linn.) Less. in DC., Prodr. 5:451. 1836; Hook. *f.*, Fl. Brit. Ind. 3:272. 1881. *Baccharis indica* Linn., Sp. Pl. 861. 1753.

Erect, pubescent undershrubs. Leaves obovate, with a tapering, base entire, acute or mucronate at the apex, acutely serrate-dentate, alternate, subglabrous, very aromatic. Flowers in terminal and axillary corymbs; bracts imbricate, 6-7 seriate, obtuse; corolla tubular, violet, in the marginal flowers filiform, shortly 3-4 dentate, disc flowers funnel-shaped, with a 5-lobed limb; anthers with a sagittate base with an obtuse top, on either side with an acute, basal appendage; in female flower style exceeding the corolla, with 2 thin branches; in male flower style entire or shortly 2-dentate. Achenes small, subturbinate.

Fl. Feb. - May; *Fr.* May - Oct.

Distr. India : Nicobar Islands, Sunderbans.

Malaysia, China.

Notes. Rare in coastal areas, on clayish, hinterlands of tidal swamps.

7. AVICENNIACEAE

Avicennia Linn.

Key to the species

1a. Leaf blades ovate or lanceolate to lanceolate-oblong or elliptic, usually abruptly acute at the apex; expanded corolla 5-8 mm wide .... **1. *A. marina***

1b. Leaf blades obovate or broadly oblong, rounded at the apex; expanded corolla 12 15 mm wide .... **2. *A. officinalis***

1. Avicennia marina (Forsk.) Vierh., Denkschr. Kaiserl. Akad. Wissensch. Wien Math.-Naturwiss. Kl. 71:435. 1907. *Sceura marina* Forsk., Fl. Aegypt.-Arab. 2:37. 1775.

Shrubs or small trees, up to 4 m tall. Leaf-blades ovate or lanceolate to lanceolate-oblong or elliptic, usually abruptly acute at the apex, glabrous. Flowers sessile in terminal cymes; calyx deeply lobed; corolla pale yellow, 5-8 mm wide, ovary villous; stigmas 2. Capsules 12-15 mm long, ovoid, apiculate, compressed.

Fl. April - June; *Fr.* June - August.

Distr. India : Nicobar Islands, Sunderbans, East & West Coast Mangrove areas.

E. Africa, Egypt, Arabia, Pakistan, Sri Lanka, S.E. Asia, China and Australia.

Notes. This species forms pure stands in the proximal zone of the estuarine mangrove swamps. In Andaman - Nicobar Islands the salinity of mangrove soil has been found to be up to 46%. This species can grow in such highly saline soils.

2. **Avicennia officinalis** Linn., Sp. Pl. 110. 1753; Parkinson, For. Fl. Andam. 218. 1923.

Trees, up to 20 m tall. Stem bark smooth, whitish-grey. Pneumatophores many, simple or forked. Leaves obovate or broadly oblong, rounded at the apex. Flowers in trichotomous sessile heads; corolla yellow, 12-15 mm wide; ovary shortly hairy. Capsules ovoid, compressed, beaked at apex.

Fl. June - August; *Fr.* August - Oct.

Distr. India : Andaman Islands, East & West Coast Mangrove areas.

Bangladesh, Myanmar, Sri Lanka, Malaysia, Java, China and in the Islands of Pacific and Indian oceans.

Notes. Sporadic in salt marshes and tidal forests at river mouth and along seashores.

8. BIGNONIACEAE

Dolichandrone Fenzl. *ex* Seeman

Dolichandrone spathacea (Linn.*f.*) K. Schum., Fl. Kais.-Wilhelm Land. 123. 1889. *Bignonia spathacea* Linn.*f.*, Suppl. Pl. 283. 1781. *Dolichandrone rheedii* Seem. in J. Bot. 3:280. 1870; Parkinson, For Fl. Andam. 215. 1923.

Trees, 3-15 m tall. Leaves imparipinnate, obovate-oblong to elliptic in outline; leaflets 5-9, ovate-lanceolate, caudate-acuminate at apex, unequal at base, glabrous above, hairy below. Inflorescences terminal racemes, 2-8 flowered; calyx spathaceous, beaked; corolla 5-lobed, white; style exserted.

Capsules with flattened pseudoseptum, semiwoody, bluntly pointed, smooth or obscurely ribbed.

Fl. May - June; *Fr.* June - August.

Distr. India : Andaman & Nicobar Islands, Malabar Coast, Sunderbans, Mahanadi Delta.

Tropical S.E. Asia, extending through Malaysia, New Caledonia, New Guinea and Micronesia.

Notes. This species is rather sporadic and confined in nearly pure stands along the edge of lagoons or scattered near the inner edge of mangrove swamps in association with *Sonneratia caseolaris* and *Heritiera littoralis.*

9. CAESALPINIACEAE

Caesalpinia Linn.

Key to the species

1a. Leaflets 6-10 pairs per pinna; flowers unisexual; pods prickly; seeds grey .... **1. *C. bonduc***

1b. Leaflets 2-5 pairs per pinna; flowers bisexual; pods not prickly; seeds black .... **2. *C. crista***

1. Caesalpinia bonduc (Linn.) Roxb., Fl. Ind. 2:362. 1832 *emend.* Dandy & Exell in J. Bot. 76.179. 1938; Hatting in Reinwardtia 9:17. 1974, *non sensu* Baker in Hook.*f.*, Fl. Brit. India 2:255. 1878. *Guilandina bonduc* Linn., Sp. Pl. 381. 1753. *Caesalpinia bonducella* (Linn.) Fleming in Asiat. Res. 11. 159. 1810; Baker in Hook.*f.*, Fl. Brit. India 2:254. 1878; Parkinson, For. Fl. Andam. 157. 1923.

Climbers, up to 10 m. Leaves pinnate or bipinnate, 3-5 leaflets with recurved prickles at the base of the pinnae; leaflets opposite or subopposite, 12-20 in all per pinna, elliptic-oblong. Racemes supra-axillary. Flowers in many flowered racemes; sepals 5, often with glandular hairs on their margins; petals 5, claw densely hairy on both sides; stamens 10; anthers hairy; filaments hairy in the basal part; pistil hairy; ovary densely set with long spines; ovules hairy; stigma ciliate. Pods thickest at the top, more or less ligneous, dehiscent. Seeds ovoid, smooth, grey.

Fl. & Fr. Throughout the year.

Distr. India : Andaman & Nicobar Islands, East & West Coast Mangrove areas.

Tropical coasts all over the World.

Notes. The ecological amplitude of this pantropic species seems to be very wide. It is often coastal, it may also occur in Inland and also grows in association of mangrove species.

2. Caesalpinia crista Linn., Sp. Pl. 380. 1753. *Guilandina nuga* Linn., Sp. Pl. ed. 2. 546. 1762. *Caesalpinia nuga* (Linn.) Ait., Hort. Kew. ed. 2. 3:32. 1811; Parkinson, For. Fl. Andam. 157. 1923.

Liana, up to 15 m, armed with recurved prickles. Pinnae 2-5 pairs, often armed; leaflets opposite, 2-5 pairs, blade coriaceous. Racemes either axillary or aggregated into terminal inflorescenes. Flowers bisexual, bright yellow dashed with red; sepals 5, lowest one cucullate, glabrous, others ciliate; petals 5, claw hairy, limb suborbicular; stamens 10, with woolly filaments. Pods indehiscent, falcate-elliptic; base cuneate, apex obtuse to acute, beaked, with 1(-2) seed. Seeds ovoid, black.

Fl. & Fr. Throughout the year.

Distr. India : Andaman & Nicobar Islands, East and West Coast Mangrove areas.

Southeast Asia, Australia, Malaysia.

Notes. Common on river banks, on sandy beaches, in and around mangroves (but then only sandy areas).

10. COMBRETACEAE

Lumnitzera Willd.

Key to the species

1a. Plants with knee shaped pneumatophores; inflorescences terminal; flowers pedicellate; petals red; stamens twice as long as the petals .... **1. *L. littorea***

1b. Plants without pneumatophores; inflorescences axillary; flowers sessile; petals white; stamens approximately equalling the petals .... **2. *L. racemosa***

1. Lumnitzera littorea (Jack) Voigt, Hort. Subarb. Calc. 39. 1845. *Pyrrhanthus littoreus* Jack, Mal. Misc. 2. 7:57. 1822. *Lumnitzera coccinea* W. & A., Prodr. 316. 1834; Parkinson, For. Fl. Andam. 168. 1923.

Trees, 8-20 m tall, with slender knee-shaped pneumatophores. Leaves narrowly obovate-elliptic, rounded at apex, cuneate at base. Flowers in terminal racemes, red, erect; calyx 5-lobed, broadly ovate; petals 5, free, glabrous, oblong-elliptic; stamens usually 10, inserted on the inner rim of the calyx cup; anthers versatile, sometimes with a minute apical appendage; ovary inferior; style simple, filiform. Fruits ellipsoid-oblong, longitudinally ribbed.

Fl. Jan. - April; *Fr.* April - Oct.

Distr. India : Andaman & Nicobar Islands.

Tropical Asia, Northern Australia, Polynesia and throughout Malaysia.

Notes. This species is always restricted to the middle zone of the mangrove forests, where soluble salts are more.

2. Lumnitzera racemosa Willd., Neue Schr. Ges. Naturf. Fr. Berl. 4:187. 1803; Parkinson, For. Fl. Andam. 168. 1923.

Shrubs or small trees, up to 8 m tall. Leaves oblanceolate or obovate, sessile, emarginate at apex, tapering at base. Flowers in axillary spikes, white, sessile; calyx 5-lobed, ovate-acuminate; petals 5, free, elliptic or oblanceolate; stamens 10. Fruits ellipsoid, woody, crowned with persistent calyx lobes.

Fl. Jan. - April; *Fr.* April - July.

Distr. India : Andaman & Nicobar Islands and along both East and West Coast from Sunderbans downwards and up to Maharashtra.

Eastern tropical Africa, Madagascar, Tropical Asia, Northern Australia, Polynesia and throughout Malaysia.

Notes. Gregarious in the muddy or sandy elevated areas in the estuarine and backwater mangroves.

11. CYPERACEAE

Fimbristylis Vahl

Fimbristylis ferruginea (Linn.) Vahl, En. 2:291. 1806. *Scirpus ferrugineus* Linn., Sp. Pl. 50. 1753.

Perennial with shortly creeping herb. Rhizome woody. Stems forming tufts, rigid. Cauline leaves much shorter than the stems, erect, conduplicate or triquetrous. Inflorescences simple or subcompound, involucral bracts 2-3, margin scarious at base. Spikelets solitary, ovoid to oblong-ovoid, terete, acute; rachilla narrowly winged; glumes spiral, ovate to oblong, obtuse, scarcely keeled, apiculate; keel 1-nerved; stamens 3; anthers oblong-linear; style flat, slightly dilated at the base, densely ciliate. Nuts 2 mm long, biconvex, obovate or oblong-obovate, shortly stipitate, straw coloured.

Fl. July - Sept.; *Fr.* Sept. - Dec.

Distr. Throughout India.

Tropics and subtropics of the World.

Notes. This species occurs in sunny wet localities specially those subject to the influence of brackish water and also found on hinterland clearings in the vicinity of mangrove forests.

12. EUPHORBIACEAE

Excoecaria Linn.

Excoecaria agallocha Linn., Syst. ed. 10, 2:1288. 1759; Parkinson, For. Fl. Andam. 242. 1923.

Trees, 3-8 m tall, with milky latex. Bark gray, lenticellate. Leaves spirally arranged, simple, coriaceous or somewhat fleshy, ovate, ovate-elliptic or ovate-oblong, apex acute or obtuse, narrowed at base, margin entire to shallowly crenulate with a minute gland in the notch in young leaves, basal glands usually circular and often associated with additional outgrowths like the marginal glands. Stipules minute, ephemeral as lateral triangular scales on each side of petiole. Inflorescences axillary; flowers unisexual, fragrant; male flowers almost sessile, yellow; tepals 3, narrow, laciniate; stamens 3, yellow; anthers bilocular, basifixed; pistillode absent; female flowers in axillary few flowered racemes, sessile, with a pair of basal bracteoles; tepals 3, wider than those in male flowers; staminodes absent; ovary trilocular. Capsules dehiscing into 3 cocci to release the solitary seed, depressed globose. Seeds black.

Fl. March - June; *Fr.* June - August.

Distr. India : Andaman & Nicobar Islands and both the East & West Coast from Sunderbans to South Narmada.

Sri Lanka, Myanmar, Tropical S.E. Asia, Malaysia, N. Australia and New Caledonia.

Notes. Common in the intertidal forests, often extending into muddy of sandy seashores.

13. FABACEAE

Key to the genera

1a. Calyx united; stamens in 1 or 2 bundles **2**

1b. Calyx divided; stamens free.. **1. *Cynometra***

2a. Leaflets opposite; pods distinctly winged .... **3. *Derris***

2b. Leaflets alternate; pods not winged .... **2. *Dalbergia***

1. Cynometra Linn.

Key to the species

1a. Sepals curved distally when reflexed at anthesis; style bent, oblique to the dorsal line of the ovary, becoming a prominent lateral beak in fruit; ovary wall hairy inside .... **1. *C. iripa***

1b. Sepals not curved distally when reflexed at anthesis; style straight, in line with the dorsal part of the ovary; fruits without a prominent lateral beak; ovary wall glabrous inside .... **2. *C. ramiflora***

1. **Cynometra iripa** Kosteletzky, Allg. med.-pharm. Fl. 4:1341. 1835. *C. ramiflora* Linn. var. *mimosoides* Baker in Hook. *f.*, Fl. Brit. Ind. 2:267. 1878; Parkinson, For. Fl. Andam. 154. 1923.

Trees, 3-7 m tall, with spreading crown or bush. Leaves with 2 pairs of leaflets; leaflets chartaceous to subcoriaceous, obliquely obovate-oblong to oblong, obtuse or very obscurely shortly broadly acuminate, base contracted into the very petiole. Inflorescences axillary pseudo-racemes, few flowered; flowers forming a globose cone; calyx segments 4-5, curved distally when reflexed at anthesis; stamens 10; ovary wall glabrous inside. Pods ellipsoid, deeply rugose, with a lateral beak like stubby nose, 1-seeded.

Fl. Sept. - Nov.; *Fr.* Dec. - Feb.

Distr. India : Andaman & Nicobar Islands, Sunderbans, Mahanadi delta.

Sri Lanka, Malaysia and North Australia.

Notes. Frequent to back of mangrove forest, in the *Heritiera littoralis* zone.

2. **Cynometra ramiflora** Linn., Sp. Pl. 382. 1758. *C. ramiflora* Linn. var. *heterophylla* Thwaites, Enum. Pl. Zeyl. 97. 1859; Parkinson, For. Fl. Andam. 154. 1923.

Trees, up to 20 m tall. Leaves with 1-2 pairs of leaflets; leaflets orbicular to ellpitic to oblong-lanceolate, apex acute to acuminate, base contracted into a very short petiolule. Raceme axillary; calyx deeply 4 lobed, reflexed, unequal, glabrous to hairy; petals 5, free, lanceolate; stamens 10;

ovary densely hairy, shortly stipitate. Pods ovate-elliptic, with shallowly rugose to undulated surface and subterminal beak, 1-seeded.

Fl. Sept. - Nov.; *Fr.* Dec. - Feb.

Distr. India : Andaman & Nicobar Islands, Sunderbans, Mahanadi delta.

Malaysia, Sri Lanka, Phillippines and North Australia.

Notes. Frequent along intertidal areas in the mangrove forests in association with *Heritiera littoralis*. Sometimes also occurs in Inland.

2. **Dalbergia** Linn. *f.*

Dalbergia candenatensis (Dennst.) Prain in Journ. Asiat. Soc. Beng. 70:49. 1901. *Cassia candenatensis* Dennst., Sch. Zum. Hort. Malabr. 12.1818. *Dalbergia monosperma* Dalz. in Hook., Journ. Bot. 2:36. 1850; Parkinson, For. Fl. Andam. 149. 1923.

Unarmed climbers. Leaves alternate, imparipinnate; leaflets 5, rarely 3 or 7, obovate to ovate-oblong, obtuse at apex, cuneate at base. Inflorescences axillary; flowers in unilateral racemes, pedicellate; bracts ovate; bracteoles 2, ovate to ovate-orbicular; calyx campanulate, 5-toothed; corolla vexillum, ovate to ovate-oblong, reflexed, long clawed; stamens 10, monadelphous; ovary long stipitate. Pods flat, indehiscent, mucronate at apex, glabrous.

Fl. May - Sept.; *Fr.* June - Nov.

Distr. India : Andaman Islands, Mahanadi, Sunderbans, East & West Coast Mangrove areas.

Sri Lanka, Burma, Bangladesh, Malaysia, North Australia and China.

Notes. This is a littoral climber; often found in the tidal forests.

3. **Derris** Lour.

Derris trifoliata Lour., Fl. Cochinch. 2:433. 1790. *Robinia uliginosa* Willd., Sp. Pl. 3:1133. 1802. *Derris uliginosa* (Willd.) Benth. in Miq., Pl. Jungh. 1:252. 1852; Parkinson, For. Fl. Andam. 147. 1923.

Scandent shrubs to climbers. Leaves imparipinnate; leaflets 3-5, ovate-oblong, rounded at base, shortly acuminate, subcoriaceous. Flowers in axillary racemes; calyx truncate; corolla rose-pink, vexillum ovate to ovate-orbicular; ovary narrowly oblong. Pods glabrous, flat, distinctly winged on the dorsal suture, reticulately veined. Seeds ovate-orbicular.

Fl. April - July; *Fr.* August - Oct.

Distr. India : Andaman & Nicobar Islands, Peninsula.

Myanmar, Sri Lanka, Bangladesh, Malaysia, N. Australia, Vietnam, Formosa, China, E. Africa, Madagascar.

Notes. Common in the mangrove swamps and creeks and also in tidal forests.

14. FLAGELLARIACEAE

Flagellaria Linn.

Flagellaria indica Linn., Sp. Pl. 333. 1753; Parkinson, For. Fl. Andam. 259. 1923.

Climbers, 2-10 m long. Leaves bifarious, ovate to lanceolate or linear, ending in a fine slender spiral tendril, base rounded or slightly cordate, abruptly contracted into a dorsally flattened petiole. Panicle erect, mostly forked. Flowers solitary, subtended by a short, broad bract; tepals oval, white. Drupes subglobose.

Fl. & Fr. Throughout the year.

Distr. India : Andaman & Nicobar Islands and throughout the East & West coast Mangrove areas.

Africa, Sri Lanka, Tropical S.E. Asia, Malaysia, Polynesia to Australia.

Notes. Common in forests along fringes of the mangroves.

15. HIPPOCRATEACEAE

Salacia Linn.

Salacia chinensis Linn., Mant. 2:293. 1767. *S. prinoides* DC., Prodr. 1:571. 1824; Parkinson, For. Fl. Andam. 127. 1923. *S. latifolia* Wall. *ex* Lawson in Hook. *f.*, Fl. Brit. Ind. 1:629. 1875; Parkinson, For. Fl. Andam. 128. 1923.

Climbing shrubs. Leaves decussate, ovate, broad-elliptic, elliptic to elliptic-lanceolate, obovate, acute at apex, base cuneate. Bracts triangular. Flowers in fascicles on axillary bracteate tubercles; calyx deeply 5-lobed, lobes triangular, semi-orbicular, obtuse or rounded; petals broad-elliptic, ovate, obovate, obtuse, yellowish or yellowish green; disc annular pulvinate; stamens 3; anthers obtuse, transversely dehiscent; pistil triangular; ovary 3-celled; ovules 2 in each cell, attached at the top of the axis. Fruits globose, 1-seeded. Seeds globose.

Fl. Jan. - March; *Fr.* March - May.

Distr. India : Andaman Islands, East and West Coast Mangrove areas.

Sri Lanka, Myanmar, Thailand, Indo-China, throughout Malaysia, North Queensland, New Britain, Solomon Island and Fiji.

Notes. Rare in forests along the seashores, hinterlands of tidal swamps.

16. LECYTHIDACEAE

Barringtonia Forst.

Barringtonia racemosa (Linn.) Spreng., Syst. Veg. 3:127. 1826; Parkinson, For. Fl. Andam. 174. 1923. *Eugenia racemosa* Linn., Sp. Pl. 471. 1753.

Small trees, 3-10 m tall. Leaves tufted at the ends of stout twigs, subsessile, obovate, tapering at base, glossy. Racemes long, pendulous, mostly terminal; bracts triangular; calyx tube rupturing into 2-5 unequal lobes; petals 4, oblong or oblong-oval, pink to red in colour; ovary 2-4 celled. Fruits ovoid, four angled, with persistent calyx.

Fl. Feb. - Sept., *Fr.* April - Oct.

Distr. India : Andaman & Nicobar Islands and throughout the East and West Coast Mangrove areas.

South East Asia, Polynesia.

Notes. Common on the shores of backwaters, estuaries and rivers, but also in Inland.

17. LYTHRACEAE

Pemphis Forst.

Pemphis acidula J.R. & G. Forst., Char. Gen. Pl. 68, t. 34. 1776; Parkinson, For. Fl. Andam. 178. 1923.

Erect or sprawling shrubs with numerous branchlets, forming a dense crown. Leaves decussate, elliptic-oblong, acute at base, thin-fleshy, densely and finely appressed-canescent on both sides. Flowers axillary, actinomorphic, 6-merous, heterostylous; calyx bibracteolate, campanulate-tubular; lobes broadly triangular; petals elliptic-oblong, white; stamens 12, alternately longer and shorter, longer ones oppositisepalous; anthers dorsifixed; ovary globose, 1-celled; style filiform; stigma broad. Capsules obovoid or ellipsoid, operculate. Seeds numerous, cuneate, alate.

Fl. August - Dec.; *Fr.* Dec. - April.

Distr. India : Andaman Islands and Southern coast of both peninsulas.

Tropical East Africa and Tropical Asia to Tropical Australia and Polynesia.

Notes. Common along edges of mangrove forests.

18. MALVACEAE

Hibiscus Medik.

Hibiscus tiliaceus Linn., Sp. Pl. 694. 1753; Parkinson, For. Fl. Andam. 96. 1923.

Trees, up to 15 m tall. Leaves orbicular, unlobed, apex acute to acuminate, base deeply cordate, chartaceous to coriaceous, 1-5 linear nectaries on the nerves beneath the lamina. Stipules ovate to oblong. Flowers solitary, axillary or in terminal raceme by reduction of upper leaves; epicalyx cupular, segments deltoid; calyx campanulate; petals obovate, yellow with purple base, apex rounded, base fleshy. Capsules globose to ovoid with a short beak. Seeds blackish brown, papillose.

Fl. & Fr. Throughout the year.

Distr. India : Andaman & Nicobar Islands, East & West Coast Mangrove areas.

Tropical and Subtropical regions of both old and new World.

Notes. Common on banks of tidal rivers in hinterland.

19. MELIACEAE

Key to the genera

1a. Leaves imparipinnate, leaflets 5-11; seeds with fleshy aril .... **1. *Aglaia***

1b. Leaves paripinnate, leaflets 1-3 jugate; seeds without aril .... **2. *Xylocarpus***

1. Aglaia Lour.

Aglaia cucullata (Roxb.) Pellegrin in Lecomte, Fl. Gen. Indo-China 1:771. 1911. *Amoora cucullata* Roxb., Pl. Corom. 3:54. t. 258. 1820; Hiern in Hook. *f.*, Fl. Brit. India 1:561. 1875.

Trees, up to 10 m tall, with numerous, vertical blind-rootsuckers or woody pneumatophores. Leaves spirally arranged, imparipinnate; leaflets 5-11, ovate-oblong-lanceolate, often falcate, unequal sided, apex obtuse, oblique at base. Male inflorescences paniculiform, female racemiform; flowers unisexual; calyx 3-lobed, lepidote, minutely ciliate; petals 3, yellow, elliptic; staminal tube petuliferous, obovoid; anthers 6, pubescent; ovary ovoid-trigonous, densely stellate-scaly. Fruits globose, trilocular capsules. Seeds covered with fleshy orange aril.

Fl. & Fr. Throughout the year.

Distr. India : Andaman Islands, Sunderbans, Mahanadi delta.

Myanmar, Java and Borneo.

Notes. Rarely distributed in the intertidal mangrove areas. Reported for Andamans by Hooker *f.* (*l.c.*).

2. **Xylocarpus** Koenig

Key to the species

1a. Buttresses present; staminal teeth prominent 2

1b. Buttresses absent; staminal teeth obscure 3

2a. Stem smooth; leaflets obovate; fruits 20-35 cm in diameter .... ***2. X. granatum***

2b. Stem rough and fissured; leaflets elliptic-oblong; fruits 7-13 cm in diameter .... ***1. X. gangeticus***

3a. Leaflets ovate; stigma cupular; fruits 5-7 cm in diameter .... ***4. X. moluccensis***

3b. Leaflets elliptic-oblong; stigma discoid; fruits 8-15 cm in diameter.......... ***3. X. mekongensis***

1. **Xylocarpus gangeticus** (Prain) C.E. Parkinson in Ind. For. 60. 2:140. 1934. *Carapa moluccensis* Lamk. var. *gangetica* Prain in Rec. Bot. Surv. Ind. 2:292. 1903. *Carapa* sp. No. 116, Parkinson, For. Fl. Andam. 118. 1923.

Trees, 10-15 m tall, small buttressed. Stem base with pointed pneumatophores. Leaves with two pairs of leaflets; leaflets elliptic-oblong, apex rounded. Flowers in axillary thyrses, white with a reddish gland inside; calyx 4-lobed; petals 4, free. Fruits globose, obscurely 4-lobed.

Fl. Feb. - May; *Fr.* May - August.

Distr. India : Andaman Islands, Sunderbans, Orissa.

Myanmar and Malaysia.

Notes. This species is included on the basis of Parkinson's report (*l.c.*). No further collections are available. It is reported to grow in tidal forests.

2. **Xylocarpus granatum** Koenig. in Naturf. 20:2. 1784. *Carapa obovata* Blume, Bijdr. 179. 1825; Parkinson, For. Fl. Andam. 118. 1923.

Trees, 5-12 m tall, buttressed. Leaves unijugate or bijugate; leaflets obovate or obovate-oblong, entire, rounded at apex, base narrowed. Flowers as in *X. gangeticus*. Fruits globose, splitting into 4-valves. Seeds pyramid shaped.

Fl. & Fr. Throughout the year.

Distr. India : Andaman & Nicobar Islands and along the East and West coast Mangrove areas.

Africa to North Australia through Sri Lanka, Malayan Archipelago.

Notes. This species is found in association with *Kandelia candel*, *Rhizophora* sp. and *Sonneratia alba*. Common on sheltered banks in the mangrove forests and forms one of the important constituents of the mangrove forests.

3. **Xylocarpus mekongensis** Pierre, Fl. Cochinch. t. 359. B. 1897; Banerjee *et al* in Mangroves in India Identification Manual, Calcutta. 60. 1989.

Trees, upto 15 m tall, trunk *ca* 50 mm diam. at base; Bark reddish brown, peeling off into thick flakes; stem base with woody pneumatophores or blind root-suckers. Leaves paripinnate, mostly bijugate; leaflets 2-4 paired, each 9-18 x 4-7 cm, elliptic-oblong, obtuse at both ends. Flowers white, in axillary panicles. Fruits 8-12 cm across, brown, globose, capsular. Seeds tetrahedral.

Fl. April - June; *Fr.* June - August.

Distr. India : Andaman & Nicobar Islands, Sunderbans, Mahanadi delta.

Malaysia.

Notes. Description of this species is based on Banerjee *et al* (*l.c.*) who reported this from Andaman & Nicobar Islands. No further collections are available. It was reported to grow in the interior elevated areas of mangrove forests in association with *Bruguiera gymnorrhiza*.

4. Xylocarpus moluccensis (Lamk.) Roem., Syn. Hessper 1:124. 1846. *Carapa moluccensis* Lamk., Encycl. 1:621. 1785; Parkinson, For. Fl. Andam. 118. 1923.

Small trees, 5-8 m tall. Stem base with woody pneumatophores. Leaves with 2 or 3 pairs of leaflets; leaflets ovate, acute at apex, oblique at base, dark green above, coriaceous. Flowers white, with red gland inside. Fruits globose.

Fl. & Fr. Throughout the year.

Distr. India : Andaman & Nicobar Islands, Mahanadi delta.

Tropical Africa to North Australia through India and Malaysia.

Notes. Common along beaches and apparently favouring open coast.

20. MYRSINACEAE

Key to the genera

1a. Leaves with pellucid colour dots; corolla reddish violet; fruits globose.. **2. *Ardisia***

1b. Leaves without any dots; corolla white; fruits elongated .... **1. *Aegiceras***

1. Aegiceras Gaertner

Aegiceras corniculatum (Linn.) Blanco, Fl. Fillip. 1:79. 1937. *Rhizophora corniculata* Linn., Diss. Herb. Ambo in. 1754. *Aegiceras majus* Gaertn., Fruct. 1:216. t. 46. 1788; Parkinson, For. Fl. Andam. 194. 1923.

Trees 5-6 m tall. Leaves ovate-oblong or obovate, coriaceous, retuse at apex, cuneate at base. Flowers axillary or in leaf opposed umbels; corolla white. Fruits curved or falcate, tapering to a pointed apex.

Fl. & Fr. March - Sept.

Distr. India : Andaman Islands, Saurashtra, Kutch to Sunderbans and almost all mangrove areas in the Tropics and Subtropics of old and new World.

Notes. Common along the middle zone of the intertidal banks of the creeks and found in association with the species of *Rhizophora*.

2. **Ardisia** Sw.

Ardisia solanacea Roxburgh, Pl. Corom. 1:27, t. 27. 1795. *A. humilis* Vahl., *sensu* C.B. Clarke in Hook. *f.*, Fl. Brit. Ind. 3:529. 1882; Parkinson, For. Fl. Andam. 194. 1928.

Small trees, 3-8 m tall. Leaves obovate-narrowly oblong, narrowed at the base, succulent-coriaceous. Flowers in axillary umbellate racemes, 5-merous; calyx-segments distinctly dotted, shortly ciliolate; corolla more or less fleshy, deeply divided, reddish-violet; stamens free; filaments very short; anthers erect, acute, conspicuously transversely plicate; ovary globose; style filiform. Drupe depressed-globose, blackish red. Seed 1, sub-globose.

Fl. March - July; *Fr.* August - Oct.

Distr. India : Andaman & Nicobar Islands and throughout East & West Coast Mangrove areas.

Sri Lanka, Singapore, Malaysia and China.

Notes. Common in mangrove swamps, sand banks of tidal forests.

21. PANDANACEAE

Pandanus Linn. *f.*

Pandanus odoratissimus Linn. *f.*, Suppl. Pl. 64. 1781.

Di- or trichotomously branched large shrubs. Leaves trifariously spiralled, gathered in terminal crowns; new branch subtended by a deltoid bifacial-biconcave prophyll; normal adult leaves sheathing at base, linear-ensiform or lanceolate or rarely narrowly elliptic, leaf margins and dorsal midrib armed with prickles. Inflorescences unisexual; staminate inflorescence a raceme of spikes, ephemeral, bracteate; stamens on short filaments; anthers narrowly sagittate; pistillate inflorescences terminal, pedunculate, trigonal, bracteate; lower bracts leaf like; upper bracts whitish-yellow, the midrib and margins with reduced prickles. Carpels each terminating in a single U- or V-shaped stigma.

Fl. Sept. - Nov.

Distr. India : Andaman & Nicobar Islands and East & West Coast Mangrove areas.

Indo-malayan and Polynesian extending to Southern China and Tropical Australia.

Notes. This species is a littoral shrub; often found in the tidal forests.

22. PLUMBAGINACEAE

Aegialitis R. Br.

Aegialitis rotundifolia Roxb., Fl. Ind. 2:111. 1832; Parkinson, For. Fl. Andam. 193. 1923.

Shrubs, up to 4 m tall. Stems straight with leaf-scars, branching only towards apex, stem base swollen with numerous stilt roots. Leaves with stem clasping petioles, alternate, orbicular, truncate or slightly apiculate at apex, rounded or broad-cuneate to slightly cordate at base. Calyx 5-lobed; corolla white, united at base forming a short tube with staminal bases; stamens 5. Capsules linear.

Fl. Feb. - April; *Fr.* April - June.

Distr. India : Andaman & Nicobar Islands, Sunderbans, Mahanadi delta.

Myanmar, Malaysian Islands.

Notes. This species was reported by Parkinson (*l.c.*). No further collections are available. It was reported as a low muddy mangrove.

23. RHIZOPHORACEAE

Key to the genera

1a. Trees with extensively developed stilt roots; flowers tetramerous; calyx 4-lobed; petals 4, entire, without distal appendages; anthers sessile, multilocellate, dehiscing by an adaxial valve .... **4. *Rhizophora***

1b. Trees without stilt roots; flowers penta to more (upto 16) merous; calyx 5 or more lobed; petals 5 or more, deeply emarginate, with apical appendages; anthers with filaments, 4-locular, dehiscing by slits.. **2**

2a. Trees with knee like pneumatophores; petals with marginal hairs; stamens twice as many as calyx lobes; hypocotyl of seedling blunt or abruptly pointed distally **3**

2b. Trees without pneumatophores; petals without marginal hairs; stamens numerous; hypocotyl of seedling slender, gradually narrowed and distantly pointed............ .... **3. *Kandelia***

3a. Leaves usually less than 10 cm long, apex rounded; calyx 5 or 6-lobed; lobes blunt, 2-4 mm long; petals less than 0.5 cm long; stamens 10 or 12 .... **2. *Ceriops***

3b. Leaves usually more than 10 cm long, apex acute; calyx 8-16 lobed; lobes pointed, 10-15 mm long; petals 0.5 - 2 cm long; stamens 16 - 32 .... **1. *Bruguiera***

1. **Bruguiera** Lamarck

Key to the species

1a. Flowers (2-) 3 (-5) in pedunculate cymes; less than 2 cm long........................**2**

1b. Flowers solitary, 3-4 cm long..**3**

2a. Calyx tube smooth; lobes stout, long (more than 3 mm), half the length of the calyx; horizontal or reflexed in fruit .... **1. *B. cylindrica***

2b. Calyx tube ribbed; lobes slender, short (less than 3 mm), only one-fourth to one-fifth the length of the calyx; erect or at most slightly spreading in fruit .... **3. *B. parviflora***

3a. Petal lobes blunt, without filamentous appendages.......................**4. *B. sexangula***

3b. Petal lobes acute, each extended into 3 filamentous appendages....**2. *B. gymnorrhiza***

1. **Bruguiera cylindrica** (Linn.) Bl., Enum. Jav. 1:92. 1827; Dagar & Basu in Journ. Eco. Tax. Bot. 7. 3:653-654. 1985 (1986). *Rhizophora cylindrica* Linn., Sp. Pl. 433. 1753.

Trees, 10-15 m tall. Stem base buttressed, producing knee like pneumatophores. Leaves oblanceolate, rarely elliptic, apex acute, cuneate at base; petioles green in colour. Flowers in axillary pedunculate cymes; calyx tube 8-10 lobed; petals white, bilobed with a bristle in the sinus, fringed with white hairs along margin. Hypocotyl cylindric, often curved, obscurely ribbed.

Fl. March - June; *Fr.* June - August.

Distr. India : Andaman Islands, Malabar and East Coast of Peninsula.

Sri Lanka, S.E. Asia, Australia.

Notes. Gregarious on stiff clay behind *Avicennia* sp. on the sea face. Sometimes in pure stands in association with *Bruguiera gymnorrhiza* in the middle zone of the mangrove swamp.

2. **Bruguiera gymnorrhiza** (Linn.) Savigny in Lamk., Encycl. Meth. Bot. 4(2):696. 1798 as *Gymnorhiza*; Parkinson, For. Fl. Andam. 164. 1923. *Rhizophora gymnorrhiza* Linn., Sp. Pl. 443. 1753. '*gymnorhiza*'.

Trees, up to 30 m tall, with knee like pneumatophores. Leaves elliptic-oblong or ovate-lanceolate, apex acute, cuneate or rarely obtuse at base. Flowers solitary; calyx 10-16 lobed, green in buds, becoming scarlet before opening; petals whitish, turning brown, bilobed with a tuft of hairs along margin at base. Hypocotyl cigar shaped with a blunt narrowed apex, slightly ribbed.

Fl. & Fr. Throughout the year.

Distr. India : Andaman & Nicobar Islands and both East-West Coasts Mangrove areas.

Tropical South and East Africa, Madagascar, Seychelles, Sri Lanka, South East Asia, Malaysia to Australia.

Notes. This species is one of the largest trees of the mangroves, common in intertidal regions along creeks and canals of mangrove forests, usually in association with *Rhizophora apiculata* and *R. mucronata* and also forms pure stands.

3. **Bruguiera parviflora** (Roxb.) Wt. & Arn. *ex* Griff. in Trans. Medd. Phys. Soc. Calc. 8:10. 1836; Parkinson, For. Fl. Andam. 165. 1923. *Rhizophora parviflora* Roxb., Fl. Ind. 2:461. 1832.

Trees, 10-15 m tall. Stem base buttressed with knee like pneumatophores. Leaves elliptic-oblong or rarely oblanceolate, apex acute, cuneate at base. Inflorescences pedunculate cyme, 2-3 flowered. Flowers greenish white; calyx tube ribbed; petals white, bilobed, each with 3 long cilia at apex, sinus with a bristle, petal margins sparsely hairy. Hypocotyl smooth, cylindrical, slightly curved.

Fl. April - July; *Fr.* July - Sept.

Distr. India : Andaman & Nicobar Islands; both East-West Coast Mangrove areas.

S.E. Asia and Australia.

Notes. This species is common in intertidal regions of estuarine swamps, usually in association with *Bruguiera gymnorrhiza, Rhizophora apiculata* and *R. mucronata*.

4. **Bruguiera sexangula** (Lour.) in Lamk., Encyl. Meth. Bot. Suppl. 4:262. 1816; Singh *et al*, Ind. For. 113. 3:216. 1987. *Rhizophora sexangula* Lour., Fl. Cochinch. 297. 1790.

Trees, 6-15 m tall, buttressed, conspicuously lenticellate on buttresses and stilt roots. Leaves elliptic-oblong, oblong or oblanceolate, apex acute, cuneate or rarely obtuse at base. Calyx green in bud, becoming yellowish before opening, whitish within, 10-13 lobed; petals 10-13, whitish, turning brown, bilobed towards the apex, stiff, fringed with white hairs along the outer margins. Hypocotyl cigar shaped, sometimes slightly curved, obscurely grooved.

Fl. & Fr. Throughout the year.

Distr. India : Andaman Islands and on both the East & West Coasts of Peninsula.

Africa, West Asia, Tropical Asia, S.E. Asia, Australia and the Pacific.

Notes. This species was reported by Singh *et al* (*l.c.*). No further collections are available. It is reported to be mainly found along with *Bruguiera gymnorrhiza* on open rock or coral terraces.

2. Ceriops Arnold

Key to the species

1a. The apex of petal with 3 clavate appendages; stamens with long, slender filaments, much exceeding the blunt anther; hypocotyl long, up to 25 cm or more, warty throughout, terete or slightly ridged .... **2. *C. tagal***

1b. The apex of petal with a fringe of filamentous appendages; stamens with a short filament, equal or exceeding by the anther; hypocotyl not exceeding 15 cm long, warty towards apex, sharply ridged .... **1. *C. decandra***

1. **Ceriops decandra** (Griff.) Ding Hou in C.G.G.J. van Steenis, Fl. Malesiana 5:471. t. 24, f-h. 1958. *Bruguiera decandra* Griff., Trans. Med. Phys. Soc. Calc. 8:10. 1836. *Ceriops roxburghiana* Arn. in Ann. Mag. Nat. Hist. 1:364. 1838.

Shrubs or small trees, 2-6 m tall. Stems supported by many stilt roots. Leaves obovate or elliptic-oblong, apex emarginate, cuneate at base. Flowers in axillary condensed cymes, 3-5 in number; calyx deeply 5-6 lobed, ovate, acute; petals 5-6, white, with broad base, apex tipped with many cilia. Hypocotyl club shaped, sharply angular and sulcate.

Fl. & Fr. Throughout the year.

Distr. India : Andaman Islands, along the East Coast and in Kerala on the West Coast Mangrove areas.

E. Africa, Madagascar, Bangladesh, Sri Lanka, Myanmar, Thailand, Malaysia and Australia.

2. **Ceriops tagal** (Perr.) C.B. Rob. in Philipp. J. Sc. Bot. 3:306. 1908. *Rhizophora tagal* Perr. in Mem. Soc. Linn. Paris 3:138. 1824.

Ceriops candelliana Arn. in Ann. Mag. Nat. Hist. 1:364. 1838. *Nom. Illeg.*; Parkinson, For. Fl. Andam. 165. 1923.

Shrubs or trees, 1-15 m tall. Stems often with small stilt roots. Leaves obovate to obovate-oblong, apex obtuse or emarginate, cuneate at base, coriaceous, dark green above, pale below. Flowers in upper axillary condensed cymes, resinous, white; calyx 5-lobed, lobes ovate, acuminate; petals 5, oblong, with 3 clavate appendages at the tip and uncinate hairs at base; stamens 10, alternatetly longer and shorter. Fruits conical. Hypocotyl club shaped, thickened towards the apex, deeply grooved and ribbed, reddish brown.

Fl. March - July; *Fr.* July - Oct.

Distr. India : Andaman & Nicobar Islands and both the East and West Coasts Mangrove areas of Peninsula.

East Africa to Australia and the Pacific.

Notes. This is much more common than *Ceriops decandra* along the intertidal banks of mangrove and also in areas nearer to and under estuarine influence.

3. **Kandelia** Wight & Arnold

Kandelia candel (Linn.) Druce, Rep. Bot. Esch. Club Brit. Isles 1913. 3:420. 1914. *Rhizophora candel* Linn., Sp. Pl. 443. 1753. *Kandelia rheedii* W. & A., Prodr. 311. 1834. *Nom. Illeg.*; Parkinson, For. Fl. Andaman 165. 1923.

Shrubs or small trees, up to 7 m tall. Stem base buttressed with stilt roots. Leaves oblong, obtuse, base cuneate, dark green above, light brown beneath. Flowers in axillary dichotomously branched cymes, 8-9 flowered, white; calyx 5-lobed, lobes linear, acute; petals 5, each divided into capillary segments. Fruits ovoid, green. Hypocotyl clavate, pointed towards the radicle end.

Fl. April - July; *Fr.* July - Sept.

Distr. India : Andaman & Nicobar Islands and along the East and West Coasts Mangrove areas of Peninsula.

Bangladesh, Sri Lanka, Myanmar, Indo-China, China, South Japan and Malaysia.

Notes. It was reported for Andaman & Nicobar Islands by Parkinson (*l.c.*) on the basis of specimens supplied by Brandis. Reported to occur in mangrove swamp.

4. **Rhizophora** Linn.

Key to the species

1a. Flowers in 2's; peduncle *ca* 10 mm long, shorter than petiole; each flower sessile; petals glabrous; stamens 12, sessile .... **1. *R. apiculata***

1b. Flowers in 4's or higher numbers; peduncle 15-40 mm long, longer than petiole; each flower with a pedicel; petals slightly to conspicuously hairy on the margins; stamens 8 or about 16, with a short filament **2**

2a. Flowers usually in 4's; peduncle about 15 mm long, stout; stamens variable in number (8-15-) 16; petals with inconspicuous marginal hairs; styles 2-3 mm long .... **2. *R. lamarckii***

2b. Flowers 4, 8 or more per inflorescene; peduncle 25-40 mm long, slender; stamens always 8; petals with densely woolly, conspicuous marginal hairs; styles either long or absent **3**

3a. Stigmas sessile; free part of ovary emerging beyond the disc; seedling warted .... **3. *R. mucronata***

3b. Stigmas on a slender style 4-5(-6) mm long; free part of ovary enclosed by the disc; seedling smooth.. **4. *R. stylosa***

1. **Rhizophora apiculata** Blume, En. Pl. Jav. 1:91. 1827. *R. conjugata* Arn. (*non* Linn.) in Ann. Mag. Nat. Hist. 1:363. 1838; Parkinson, For. Fl. Andam. 164. 1923.

Trees, up to 30 m tall. Stems supported by numerous branched stilt roots. Leaves elliptic-oblong to sublanceolate, apex acute to apiculate, cuneate at base, coriaceous. Flowers sessile, white or cream coloured, developing in pairs from the cupular involucre; calyx 4 lobed, ovate, concave, acute, reflexed in fruit; petals 4, lanceolate, glabrous, caducous; stamens 12, 4 pairs episepalous, 4 epipetalous. Hypocotyl cylindrical, clavate.

Fl. May - July; *Fr.* July - Sept.

Distr. India : Andaman & Nicobar Islands, East & West Coasts of Peninsula.

East Africa, Sri Lanka, Bangladesh, Myanmar, South & South East Asia and Australia.

Notes. This species is gregarious along the inter-tidal regions of creeks mostly in the sheltered parts of mangrove forests, nearer to estuarine conditions.

2. **Rhizophora lamarckii** Montrouziar, Mem. Ac. Sc. Lyon 10. 1860; Singh *et al* in Ind. For. 113. 3:214. 1987.

Trees, up to 10 m tall. Stems supported by numerous, branched stilt roots. Leaves elliptic, rarely ovate. Inflorescences pedunculate, 2-flowered, rarely 3-4; calyx 4 lobed, ovate, pale yellow; petals 4, lanceolate, membranous, white, densely hairy along margin and sparsely hairy on the adaxial side; stamens 8-16, arranged in two distinct whorls. Fruits conical with exserted hypocotyl.

Distr. India : South Andaman.

New Caledonia, Bismarck, Archipelago, Solomon Islands, Sri Lanka, Australia and Malaysia.

Notes. This species was reported by Singh *et al* (*l.c.*) from Havelock Island of South Andaman, to have been found along with *Rhizophora apiculata* and *R. mucronata* on rocky substratum. No further collections are available.

3. **Rhizophora mucronata** Lamk., Encycl. 6:189. 1804; Parkinson, For. Fl. Andam. 164. 1923.

Trees, up to 25 m tall. Stems supported by numerous branched stilt roots. Leaves broad-elliptic to oblong, acute or blunt with a rolled up tip (mucro) at apex, base cuneate. Flowers in axillary cymes, 4 - 8 in number, pedicellate, cream coloured; calyx 4-lobed, ovate; petals lanceolate, densely hairy along the margins; stamens 8, 4-epipetalous, 4- episepalous, sessile. Hypocotyl cylindric, slightly rugose.

Fl. July - Sept.; *Fr.* Sept. - Nov.

Distr. India : Andaman & Nicobar Islands, Sunderbans and other East Coast Mangrove areas.

East Africa, South & South East Asia, North East Australia and Pacific isles.

Notes. This species is generally gregarious near the intertidal banks of creeks or in estuaries. It is more tolerant for sandy soil than *Rhizophora apiculata*.

4. **Rhizophora stylosa** Griff., Nat. pl. As. 4:665. 1854; Ic. 4. t. 640. 1854; Singh *et al* in Ind. For. 113. 3:214. 1987.

Trees, up to 12 m tall. Stems supported by many branching stilt roots. Leaves broadly elliptic, clustered towards end of branchlets, coriaceous, obtuse at apex, base cuneate. Flowers in dichotomously branching, axillary cymes, 8-16 in number, pedicellate, calyx 4-lobed; petals 4, lanceolate, yellowish, purple fringed at apex, densely hairy along the margins; stamens 8, 4- episepalous; styles slightly forked. Hypocotyl cylindric.

Fl. July - Sept.; *Fr.* Sept. - Nov.

Distr. India : Andaman Islands, Gahirimata estuary in Orissa coast.

Malay Peninsula, Java, Formosa and North Australia.

Notes. This species was reported for Andaman Islands by Singh *et al* (*l.c.*) from South Andaman on open coastal areas of Burmanala, Chidiatapu and Corbyn's Cove.

24. RUBIACEAE

Scyphiphora Gaertn. *f.*

Scyphiphora hydrophyllacea Gaertn.*f.*, Fruct. 3:91. t.196. 1805; Parkinson, For. Fl. Andam. 189. 1923.

Erect shrubs. Leaves opposite, oblong-obovate, acute at base, rounded or subemarginate, glossy above, dry rather thick coriaceous. Flowers in supra-axillary, glabrous cymes; calyx tube narrowly obconical, glabrous;

corolla tube narrowly obconical; lobes ovate-oblong or ovate, white; stamens inserted at throat; anthers with sagittate base, dorsifixed below the middle; disc annular, lobed; ovary 2-celled; pyrene ellipsoid-cylindric, with 8-10 longitudinal grooves, corky. Seeds oblong, terete.

Fl. & Fr. March - August.

Distr. India : Andaman Islands and along East Coast of Peninsula.

Java.

Notes. Frequent in swamps along mangrove creeks in most localities.

25. SONNERATIACEAE

Sonneratia Linn.*f.*

Key to the species

1a. Calyx cup shaped, enclosing the base of the fruit, prominently ribbed........... **1. *S. alba***

1b. Calyx flat, extended horizontally, not enclosing the fruit, at most obscurely ribbed **2**

2a. Twigs not pendulous; leaves obovate to suborbicular; veins conspicuous, prominent on the upper blade surface; petals absent; filaments white .... **3. *S. griffithii***

2b. Twigs pendulous, slender; leaves elliptic-oblong or oval-obovate to narrowly elliptic; veins inconspicuous; petals usually present; filaments red below, white distally .... **2. *S. caseolaris***

1. **Sonneratia alba** J. Smith in Rees, Cycl. 2:33. 1819; Parkinson, For. Fl. Andam. 179. 1923.

Trees, 5-8 m tall with many pneumatophores. Leaves elliptic, oblong or suborbicular, broadly rounded at apex, narrowed to a short petiole at base, succulent. Flowers usually 3-together, white; calyx 6-8 lobed, lobes ovate-oblong; petals small; ovary depressed globose; stigma capitate. Fruits with persistent calyx, obconical.

Fl. March - June; *Fr.* June - Sept.

Distr. India : Andaman Islands.

Tropical East Africa, South East Continental Asia, Madagascar, Seychelles, Myanmar, Malaysia, North Australia and Micronesia.

Notes. This species is very uncommon, grows along the mouth of tidal creeks. It prefers salt water and grows as well on sandy or rocky soil.

2. ***Sonneratia caseolaris*** (Linn.) Engl. in Engler & Prantl, Nachtr. 261. 1897, *em.* Sm. in Rees, Cycl. 33. 1819. *Rhizophora caseolaris* Linn. *p.p.* in Stickman, Herb. Amb. 13. 1754. *Sonneratia acida* Linn. *f.*, Suppl. 252. 1781; Smith in Rees, Cycl. 1:33. 1819; Parkinson, For. Fl. Andam. 178. 1923.

Small trees, up to 6 m tall, with many pneumatophores. Leaves elliptic, oblong or obovate, mucronate at apex, base cuneate. Flowers solitary, terminal, reddish purple; calyx 6-lobed, lobes longer than the tube; petals 6, linear-lanceolate, membranous; ovary depressed, globose; stigma capitate. Fruits with persistent calyx.

Fl. March - May; *Fr.* May - July.

Distr. India : Andaman & Nicobar Islands and both East and West Coast of Peninsula.

Bangladesh, South East Asia, Ceylon to North Australia, Philippines and Malaysia.

Notes. This species is very common along the intertidal banks of creeks in mangrove forests. It thrives well in less salty areas of mangrove forests on deep muddy soil, which is under fresh water inundation.

3. **Sonneratia griffithii** Kurz in J. Asiat. Soc. Beng. 40. 2:56. 1871; Parker in Ind. For. 51. 507, 510. 1925.

Trees, up to 20 m tall, with numerous pneumatophores. Leaves obovate to suborbicular, thickly coriaceous, emarginate at apex, base cuneate. Flowers apetalous, greenish white, solitary; calyx 6-8 lobed, spreading in fruit; stamens many, white. Fruits ovoid-globose, partly enclosed by persistent calyx at base.

Fl. April - July; *Fr.* July - Sept.

Distr. India : Andaman Islands, Mahanadi delta.

Africa, Bangladesh, Myanmar, West Malay Peninsula and Australia.

Notes. It was reported by Parker (*l.c.*) from Andaman Islands. It is sporadic on muddy banks of estuarine mouths.

26. STERCULIACEAE

Heritiera Aiton

Heritiera littoralis Dryand. in Ait. Hort. Kew. 3:546. 1789; Parkinson, For. Fl. Andam. 102. 1923.

Trees, up to 20 m tall, buttressed. Leaves elliptic or oblong, acute or acuminate at apex, obliquely cordate at base, upper surface green, lower surface pale, covered with fimbriate scales. Flowers axillary, in much branched small panicles, golden-yellow, tinged with red inside, stellate pubescent, unisexual; calyx connate, 4-5 dentate; androgynophore white, ringed by the anther lobes; female flowers with sterile anthers at the base of the ovary. Fruits ellipsoid, woody, smooth, shining brown, inside densely short hairy, outer side with a short radder-like apical crest.

Fl. Nov. - Jan.; *Fr.* Feb. - June.

Distr. India : Andaman Islands, Malabar, North Kanara, South Konkan coast, Mahanadi delta.

Tropical Africa, Sri Lanka, Myanmar, Thailand, Vietnam, Java, Malaysia and Australia.

Notes. Very common on sandy tidal banks in the back mangrove region.

27. TILIACEAE

Brownlowia Roxburgh

Brownlowia tersa (Linn.) Kosterm. in Council Sci. Indonesia Bull. 1:73. 1959 *et* in Reinw. 4:536. 1959. *Glabraria tersa* Linn., Mant. 2 :276. 1771. *Brownlowia lanceolata* Benth. in Journ. Linn. Soc. 5, Suppl. 2:57. 1861; Parkinson, For. Fl. Andam. 105.1923.

Shrubs, 2-3 m tall. Leaves lanceolate, dark green and glossy above, densely dull silvery scaly beneath, apex acuminate, rounded at base. Flowers in axillary or terminal cymes; calyx 4-5 lobed, connate below; petals 5, white, free; stamens many, alternating with 5 petaloid staminodes opposite the petals. Fruits pyriform, woody with a line separating the 2-valved carpels, brownish, scaly.

Fl. May - July; *Fr.* July - Sept.

Distr. India : Andaman Islands, Sunderbans, Mahanadi delta.

Myanmar.

Notes. Rarely distributed along the intertidal regions of creeks in the back mangrove areas of tidal forests. It often grows along the banks and is almost submerged during the high tides.

28. VERBENACEAE

Clerodendrum Linn.

Clerodendrum inerme (Linn.) Gaertn., Fruct. 1:271. pl. 75. 1788. *f.* **inerme** Linn., Fl. Zeyl. 231. 1747; Parkinson, For. Fl. Andam. 219. 1923.

Straggling shrubs, up to 3 m tall. Leaves decussate-opposite or rarely ternate, elliptic or narrowly lanceolate, shortly acuminate to blunt or obtuse at apex, base acute. Inflorescences axillary cymose, mostly 3-7 flowered; calyx 5-toothed, campanulate, apex acute; corolla white, hypocrateriform, tube slender, lobes obovate; stamens exserted. Fruits drupaceous, obovate.

Fl. Feb. - June; *Fr.* June - Oct.

Distr. India : Andaman & Nicobar Islands and throughout East & West Coasts Mangrove areas.

Tropical and Subtropical Asia from India to Australia. Pacific oceanica, Southern China, Brazil, Zaire, Mauritius.

Notes. This mangrove associate and slightly salt resistant species is growing in seashores, saline marshes and swamps, muddy tidal river banks and the edges of mangrove forests. It prefers moist sandy soil or sandy loam. It is seen that in many localities the mangroves genera like *Bruguiera*, *Rhizophora* have almost disappeared or/and degraded into low, seral thickets of *Clerodendrum inerme*. This species along with *Acrostichum aureum* and *Acanthus* sp. are invading ground in clearings and cuttings, specially the colossal mud heaps accumulated by mangroves crab fish and finally form dense thickets in which regeneration of the mangrove proves difficult.

This species is listed as one of the sand-stabilizing plants of deserts and semi-deserts of the World.

Chapter 4

ECONOMIC SIGNIFICANCE OF MANGROVES

The true value of mangroves, for long disregarded and abused; has increasingly been recognised by science. Mangroves are diverse and complex ecological system that contain many species. Mangroves offer a rich variety of products to people living in their vicinity, occupy a key position in food chains, provide natural protection to coastlines, perform vital cycling and filtering tasks, regulate water flows and provide a habitat for wild animals and birds.

The directly marketed goods produce by a mangrove ecosystem represent only a fraction of the total array of goods and services that yield value and benefits to individuals and societies. As a result, total values and benefits have been underestimated and conversion has become the most attractive alternative such as aquaculture ponds, infrastructure (e.g. ports, coastal roads etc.), agriculture, housing and even garbage dumps.

However, with increasing knowledge of both the econonic and environmental values of mangrove ecosytems, economic analysis can serve as a useful tool in analyzing development alternatives for mangrove forests.

Comprehensive economic analysis has shown that mangroves often yield greater social net benefits if they remain as natural ecosystems. When conversion of mangroves is clearly justified, a physical-social-economic analysis can help plan conversions that will minimize the loss of the mangrove forest benefits.

The analysis should consider the following costs of conversion (direct or indirect); possible employment generating potential benefits of conversion (large and sustainable); and a valuation of the mangroves marketed and non-marketed goods and services, both within (on-site) and outside (off-site) the area covered by mangroves (Table-4).

When the value of an existing mangrove forest is assessed, the analysis traditionally has included only those items found in quadrant 1 of Table-4, goods and services that have market prices and are found within the mangrove. The value of these products of quadrant 1 is frequently small compared to the expected benefits from conversion. Consequently, extensive mangrove areas are converted each year in the name of "economic efficiency". As knowledge has been gained of the ecosystem interactions in mangroves, greater attention has been paid to quadrant 2 effects, which occur off-site but are marketed. Goods and services of quadrants 3 and 4, found both on-site and off-site mangroves, are usually ignored. Because they are non-marketed items. For some of these goods and services, the valuation problem is due to its subsistence role in the local economy (e.g. traditional medicine, minor mangrove forest product). For others identification and quantification of the effect are sometimes a problem and placing monetary values are therefore be difficult. Nevertheless, these contributions of the mangroves may be quite important in terms of the total benefits produced by the mangrove system (Dixon, 1989). The environmental services of mangroves can be divided into products and functional. Products are the tangible output of a mangrove and can usually be used directly or sold; functions are what the mangrove environments, and their value must be inferred in indirect ways.

Table-4: Market valuation and location of mangrove goods and services

	On-Site	Off-Site
Marketed	1 Usually included (e.g. wood, charcoal, wood chips, fish, and crabs from mangroves).	2 May be included (e.g. fish from offshore).
Non-marketed	3 Seldom included (e.g. domestic fuelwood, food in times of famine, nursery, area of juvenile fish, feeding ground for estuarine fish and shrimp, medicinal uses, viewing and study of wildlife).	4 Usually ignored (e.g. nutrient flows to estuarines and buffer to storm damage).

Source : Hamilton and Snedaker (1984)

4.1 PRODUCTS OF MANGROVES

Properly managed mangroves can provide a continuing steam of products (Table-7) to local people. If the resource is over exploited, the harvest will not be sustainable, and the wider environmental functions of the mangrove may be impaired.

4.1.1. Commercial Importance

Apart from the traditional use of mangrove trees for firewood, the local people also collect wax, honey, shells, gol pata or dhani leaves (*Nypa* leaves) etc. from mangrove forests.

i) Timber :

Mangroves, typical of many tropical coastal areas, are striking examples of adaptation to shifting and unstable ground and constantly varying water levels.

In the process they develop large and complex root systems. This extensive biomass is useful for fuelwood, timber, fodder and many local construction purposes (boat, buildings, fencing, furniture etc.).

In India, including Andaman-Nicobar Islands mangrove trees are used for house building, furniture, transmission as well as telegraph poles, and certain household items (Table-5).

It has been estimated that the average yield from the North Andaman, Middle Andaman and South Andaman being 115 cu m/ha, 105 cu m/ha and 59 cu m/ha, respectively (Balachandran, 1988).

The species *Bruguiera gymnorrhiza*, *Heritiera littoralis, Rhizophora apiculata* and *R. mucronata* are the main species for extraction purpose. Other species of *Avicennia, Bruguiera, Rhizophora, Sonneratia, Xylocarpus* etc. are used for minor purposes such as posts, beams, electric poles, furnitures and agricultural implements.

ii) Charcoal :

In Thailand, the main use of mangroves is in the production of charcoal. But there is no information about charcoal production from Mangrove in Andaman-Nicobar Islands.

iii) Firewood :

Because of the high specific gravity of Rhizophoraceous wood, the species of *Bruguiera, Ceriops, Kandelia* and *Rhizophora* are preferred for firewood. Even though most mangrove species (e.g. *Avicennia, Cynometra, Heritiera, Lumnitzera, Sonneratia, Xylocarpus* etc.) are used for firewood, not all of them are useful. For example, *Rhizophora* sp. are very popular as they burn evenly with little smoke and easy to chop.

Xylocarpus sp. on the otherhand, do not give uniform heat and burn very fast. The firewood obtained from mangrove forests is either used locally or marketed in neighbouring towns or in saw mills.

Table-5: Timber producing species and their uses

Genus	*Rhizophora* *R. apiculata* *R. mucronata*	Transmission and telegraphic poles, piles, posts, beams, fences, shipbuilding, rafts, rough furnitures, firewood, flooring.
Genus	*Bruguiera* *B. cylindrica* *B. gymnorrhiza* *B. parviflora* *B. sexangula*	Piles, poles, posts, shipbuilding, rafts, firewood.
Genus	*Avicennia* *A. marina* *A. officinalis*	Piles, posts, poles, fences, wooden crafts, furnitures and used for brick burning.
Genus	*Xylocarpus* *X. granatum* *X. moluccensis*	Piles, poles, posts, fences, general construction, rafts, furnitures, beams, doors, firewood.
Genus	*Sonneratia* *S. alba* *S. caseolaris*	Piles, poles, home posts, fences, planks, furnitures, boxes, firewood, boats.
Genus	*Heritiera* *H. littoralis*	Piles, ship and boat building, beams.
Genus	*Ceriops* *C. decandra* *C. tagal*	Piles, poles, posts, ship and boat building, rafts, furnitures, firewood.
Genus	*Excoecaria* *E. agallocha*	General carpentry, matches, repairing of ships.

iv) Tannin :

As the bark of the mangrove tree has a high tannin content it is of value for tannin production. The barks of the species of *Bruguiera, Ceriops, Kandelia, Rhizophora, Sonneratia, Xylocarpus* etc. yield tannin of commercial value. Indian mangrove trees have 35% tannin in their bark, which is higher compared to other countries (Roy Chowdhury, 1964). But there are no larger tanneries in India using this resource. However, Indian fishermen of the Islands frequently use the extracts from mangrove bark to dye their fishing nets to enhance their durability.

v) Fish and shellfish :

Mangrove forests serve as a link between terrestrial and marine ecosystems. There is an import of nutrients from the land to the mangroves and export of organic matter from the mangroves to the sea (Jayewardene, 1984).

Thus at the interface between land and sea, mangroves are rich in nutrients, which provide a variety of ecological riches making for a wealth of aquatic life. They are, therefore, important for fishing, shellfish and aquaculture.

vi) Birds, animals and other wildlife :

Meat and other products (e.g. eggs, skins, honey, wax etc.) are important supplements to the diets of local people in many areas.

vii) Medicine :

Different mangrove species are used locally by the tribals of Andaman-Nicobar Islands for medicinal purposes (Chopra *et al*, 1956; Chapman, 1976; Dagar and Dagar, 1986; Dagar, 1989). Medicinal values of some mangroves are shown in Table-6.

Table-6: Mangrove plant species possessing medicinal values

Species	Part/s used	Therapeutic use
Acanthus illicifolius	Leaves	Rheumatism and Neuralgia
Acrostichum aureum	Rhizome	Wounds and boils
Barringtonia racemosa	Fruits	Cough, Asthma, Diarrhoea
Caesalpinia bonduc *C. crista*	Seeds, Leaves & Bark	Jaundice, Febrifuge & Anthelmintic
Ceriops tagal	Shoots	Malaria
Cynometra ramiflora	Leaves	Skin disease, Leprosy, Scabies and Cutaneous disease
Heritiera littoralis	Roots	Purgative
Hibiscus tiliaceus	Seeds	Diarrhoea and Dysentery
Xylocarpus granatum	Leaves	Stomach troubles
X. moluccensis	Bark	Febrifuge, Dysentery

viii) Other Products :

Honey collected from *Aegialitis rotundifolia* and *Cynometra ramiflora* has a good market value and is demand. Honey from other species like *Ceriops* sp. and *Excoecaria agallocha*, however although common is not highly valued (Blasco, 1975).

Bark and roots of *Aegiceras corniculatum* and *Derris heterophylla*, bark and fruits of *Barringtonia asiatica*, bark of *Barringtonia racemosa* and milky latex of *Excoecaria agallocha* are used as fish poisoning.

Fibres from the bark of *Pandanus* sp. and *Hibiscus tiliaceus* are used by tribals for cordage and ropes and making mats (Dagar *et al*, 1991).

Species of *Avicennia, Ceriops, Cynometra, Rhizophora, Sonneratia* are very good fodders. Goats feed on the leaves of *Cynometra* eagerly (Dagar *et al*, 1989). It is reported that Deers also feed on the leaves of the seedlings of *Rhizophora apiculata* and *R. mucronata*.

The tribes Onge prepare a kind of tea from the leaves of *Heritiera littoralis* and *Hibiscus tiliaceus* (Dagar *et al*, 1991).

Table-7: Products of Mangroves

Use	Products
Fuel	Firewood/Charcoal for cooking, smoking, heating.
Construction	Timber for buildings, boats.
Fishing	Wood for traps.
Agriculture	Green manure.
Paper production	Pulp, Fibre.
Carbohydrate foods	Sugar, Alcohol, Oil, Vinegar, Honey.
Protein foods	Finfish, Shellfish, Wildlife.
Drugs	Medicines from bark, leaves, fruits, seeds, roots and rhizomes.
Household	Furnitures, Glue, Handles, Matchsticks, Incense.
Textile/Leather	Fibres, Dye, Tannins, Leathers.
Education etc.	Wildlife, Vegetation, Habitat.

4.2 FUNCTIONS OF MANGROVES

People in the tropical coastal region have used mangrove resources for centuries without upsetting the ecological balance in the area. This was because the population involved was small. Later, over-exploitation and destruction of the mangrove areas become a consequence of population pressures compounded by unwise political decisions on mangrove allocation made in various Asian countries in the last 20 years or so. Moreover, colonial exploiters never took to the mangroves very kindly, branding them as striking, impenetrable, mosquito-infested dirty areas. There was virtually no official hesitation in reclaiming mangrove land since realization of its ecological importance only surfaced during 1960s (Rao, 1987). Even in developed countries like Australia, mangrove used to be considered as wasteland ripe for development (Hegerl, 1984).

But now, natural functions of mangroves are understood. The irreversibility of their loss makes them a prime case for caution and postponement of drastic action.

The following are the different types of their functions (Table-8).

Table-8: Environmental Services of Mangroves

Source	Product / Service
Trunks, Prop roots	Shoreline protection
Vegetation	Foodchain support
Vegetation, Soils	Nutrient/metal uptake
Habitat as Nursery	Recruitment to fisheries

4.2.1 Shoreline protection

Mangrove swamps are important in retarding coastal erosion and protecting the coast. Mangrove vegetation protects land from erosion and helps to build and consolidate silt. The mangrove swamps act as a system in which trees in all the zones have different roles to play. Damage to the system can result in irreversible coastal erosion.

Due to this function, many efforts to replant the mangroves on eroding mudflats have been tried in order to arrest the erosion. However, mangroves cannot establish below Mean Sea Level (M.S.L.). Since the levels of the eroded mudflats are generally below M.S.L., the efforts were not successful. Between M.S.L. and Mean High Water Spring (M.H.W.S.), mangroves grow very well and can provide adequate protection to the bund. The key is therefore to protect the soil beneath the mangroves, so that the trees will be able to do the rest of the work in reducing the wave energy (Othman, 1994).

4.2.2 Coastal productivity

Mangrove swamps are one of the richest and most productive areas and form the base of the food chain in the sea and coastal waters (Foo and Wong, 1980). The most important role for the mangroves is probably as a link between the terrestrial and marine ecosystems.

According to the energy model the mangrove ecosystem has two mangrove compartments (above ground structures and mud); two external energy sources (sun and upland runoff); five mangrove processes (primary productivity, above ground respiration, respiration of mud, recycling of mineral nutrients and export of organic matter outside) and six potential stress factors (channelization, drainage, siltation, hurricanes, herbicides and thermal loading) (Lugo and Snedaker, 1974).

There is an import of nutrients from the land to mangroves and export of organic matter from mangroves to the sea (Jayawardene, 1984). The floral elements in mangrove forests such as angiosperms, phytoplankton and marine algae are responsible for primary productivity. Zooplankton and benthic elements are the secondary and tertiary producers respectively (Rao, 1987). Thus mangroves provide a rich source of nutrients for all forms of life, including fish, and are favoured breeding grounds for certain kinds of marine life.

The solar radiation interacts with nutrient and mangrove biomass to yield a gross production, part of which is respired, part used to increase forest biomass and the remainder deposited as detritus.

The most important group of organisms, the penacid shrimps are dependent on the mangroves for food and shelter during the larval and juvenile stages of their life cycle. These shrimps spawn in nearshore waters and the newly hatched larvae stay along the coasts using the mangrove as a refuge area and migrate to offshore waters where they reach juvenility.

Fishing is an important occupation in and around the mangrove areas. In Andaman & Nicobar Islands fishing practices centered within or close to mangrove and estuarine areas, where individual or group efforts are localized with generally low catches.

4.2.3 Coastal aquaculture

It has been estimated that about 1.2 million hectares of mangrove areas in the Indo-Pacific region is presently used for aquaculture (Rabanall, 1977). Coastal (brackish water) aquaculture is basically the system of constructing fish traps in the form of coral and barriers to hold back fish, which have come in during the high tide.

The availability of large unpolluted water columns in protected bays and tidal creeks of Andaman & Nicobar offer ideal conditions for establishing pan culture for fishes and prawns in shallow regions and floating raft/cage culture for fishes and molluses in deeper areas (Dorairaj and Soundararajan, 1987 a, b).

It has been estimated that about 3860 ha of marshy low lying areas and mangrove swamps are found in Andaman & Nicobar Islands which can be used for brackish water and fish culture (Mongia *et al*, 1992).

Dorairaj and Soundararajan (1987 b) estimated the potential yield of fishes from Bay Islands around 1.6 lakh tons, comprising 1 lakh tons of Tunas and Tuna like fishes, 0.4 lakh tons of Plagic fish and 0.2 lakh tons of Demersol fish.

4.2.4 Habitat for wildlife

Apart from the creatures caught for local consumption, mangroves provide an important habitat for interesting animals, such as salt-water crocodiles, sea turtles, water monitor lizard and several birds.

4.2.5 Wildlife sanctuary and recreational areas

There are some species endemic to the Andaman-Nicobar Islands, while others are introduced. These are wild Goats of Barren Islands, Hornbill of Narcondam Islands, Giant robber Crab of South Sentinel Island.

To conserve of various unique and rare species of animals against extinction certain area of these Islands have been reserved as park and sanctuaries. So far, one biosphere reserve, nine national parks and 94 wildlife sanctuaries have been established in these Islands. A complete list of biosphere reserve, national parks and wild life sanctuaries are enumerated in Table-9.

Table-9: Existing National Parks and Wildlife sanctuaries in Andaman & Nicobar Islands

State	National Parks	Size km²
Andaman & Nicobar	Middle Button Island	0.64
	Mount Harriett	46.62
	North Button Island	0.44
	Saddle Peak	32.54
	South Button Island	0.03
	Mahatma Gandhi Marine	281.50
	Campbell	429.23
	Rani Jhansi Marine	256.14
	Galathea	110.00
Total	**9**	**1157.14**
	Wildlife Sanctuaries	**Size km²**
	Arial island	0.05
	Bamboo island	0.05
	Barren island	8.10
	Battimalv island	2.23
	Belle island	0.08
	Bennett island	3.46
	Bingham island	0.08
	Blister island	0.26
	Bluff island	1.14
	Bondoville island	2.55
	Brush island	0.23
	Buchanan island	9.33
	Chanel island	0.13
	Cinque island	9.51
	Clyde island	0.54
	Cone island	0.65
	Curi FW (R P) island	0.16
	Curlew island	0.03

State	Wildlife Sanctuaries	Size km²
Andaman & Nicobar	Defence island	10.49
	Dot island	0.13
	Dottrell island	0.13
	Duncan island	0.73
	East island	6.11
	East of Inglis island	3.55
	Egg island	0.05
	Elat island	9.36
	Entrance island	0.96
	Gander island	0.05
	Girjan island	0.16
	Goose island	0.01
	Hump island	0.47
	Interview island	133.87
	James island	2.10
	Jungle island	0.52
	Kwangtung island	0.57
	Kyd island	8.00
	Latouche island	0.96
	Landfall island	29.48
	Lohabarrack (Salt water croc.)	22.21
	Mangrove island	0.39
	Mask island	0.78
	Mayo island	0.10
	Megapode island	0.12
	Montogemery island	0.21
	Narcondam island	6.81
	North island	0.49
	North Brother island	0.75
	North Reef island	3.48
	Oliver island	0.16
	Orchid island	0.10
	Ox island	0.13
	Oyster island - I	0.08
	Oyster island - II	0.21
	Paget island	7.36
	Parkinson island	0.34
	Passage island	0.62
	Patrick island	0.13
	Peacock island	0.62
	Pitman island	1.37
	Point island	3.07
	Potanma island	0.16
	Ranger island	4.26

State	Wildlife Sanctuaries	Size km²
Andaman & Nicobar	Reef island	1.74
	Roper island	1.46
	Ross island	1.01
	Rowe island	0.01
	Sandy island	1.58
	Sea Serpent island	0.78
	Shark island	0.60
	Shearme island	7.85
	Sir Hugh Rose island	1.06
	Sisters island	0.36
	Snake island - I	0.73
	Snake island - II	0.03
	South Brother island	1.24
	South Reef island	1.17
	South Sentinel island	1.61
	Spike island - I	0.42
	Spike island - II	11.70
	Stoat island	0.44
	Surat island	0.31
	Swamp island	4.09
	Table (Delegarno) island	2.29
	Table (Excelsior) island	1.69
	Talabaicha island	3.21
	Temple island	1.04
	Tillongchang island	16.83
	Tree island	0.03
	Trilby island	0.96
	Tuft island	0.29
	Turtle island	0.39
	West island	6.40
	Wharf island	0.11
	White cliff island	0.47
Total	**94**	**372.13**

Wildlife sanctuaries have been created in Baren, Narcondum and South Sentinel Islands, Iobabarrack sanctuary for salt water crocodile, Sir Hugh Ross, South Reef Islands, South Sentinel Island, North Reef Islands sanctuaries for Turtles. A marine National park has been created at Wandoor for sea Turtles.

Chapter 5

CONNECTIONS BETWEEN MANGROVE, SEAGRASS AND CORAL REEF ECOSYSTEMS IN THE COASTAL ZONE

Interactions between mangrove, seagrass and coral reef ecosystems can be classified into six major types (Fig. 3).

5.1 PHYSICAL INTERACTIONS

Mangroves, seagrasses and coral reefs interact physically in a number of ways. Among these reduction of water energy, sediment relationships and flow regulations are important.

Mangroves and seagrasses are highly dependent on hydrodynamic barriers such as coral reefs, which dissipate wave energy. Mangrove and seagrass communities are enhanced in structure when these barriers are present.

Reefs are active producers of carbonate skeletal materials. These skeletal materials are constantly produced in large amount in the reef front and broken up by abrasion and ground into smaller fragments. Biological activities by fishes, urchins and boring sponges are also important in the breakdown of the calcareous materials. These processes lead to the formation of gravels, sands and silts. Some of which are transported by waves and currents into lower energy environment behind reef front. The sediments from shoals and Islands molded by waves and currents accumulate as sediment aprons, which become the reef lagoon. They eventually may be colonized by mangroves and as well as seagrasses. A living reef growing faster than it is being eroded ensures the maintenance of this barrier to wave energy and thus the presence of a lower energy environment favourable for mangrove and seagrass.

Mangroves develop in the low-lying areas bathed by saline water. The fringe and basin forests act as depositional basins and are effective binders of sediments. As a result the sediment loads into coastal waters are reduced.

Fringe and basin forests also regulate fresh water flows into coastal areas. Fresh water stored in the forest may be lost to evapotranspiration and also dissolves and dilutes excess salts accumulated in the inner fringes during the periods of drought. Thus the volume of undiluted fresh water reaching the coast is reduced.

5.2 NUTRIENTS

The inorganic nutrients phosphorus and nitrogen are essential to the primary producers of all ecosystems.

Each of the three ecosystems i.e. mangrove, seagrass and coral reef, utilize dissolved nutrients and hence reduce the concentration as water passes through the system.

Mangrove forest structure is positively correlated with high nutrient inputs; seagrasses in turn tolerate a higher level of eutrophication than coral reefs, while reefs are basically oligotrophic system least tolerant of nutrient enrichment.

Mangrove systems tend to export nutrients. Some of the excess nutrient may also be tied up as peat in system, but a sizeable fraction of the surplus may in turn nourish reef organism as dissolved and particulate organic matter.

Animals also transport nutrients between these ecosystems. For example, birds those nests in the mangroves and feed in seagrasses more nutrients into the mangroves by defecating there after returning from feeding.

5.3 DISSOLVED ORGANIC MATTER (D.O.M.)

Dissolved organic matter is a normal component of runoff from land. It is derived from leachate from the leaves, stems and roots of living plant, decomposed plant litter, soil organic matter and animal wastes. All these three ecosystems produce D.O.M.; there is a pattern of export in the direction from mangroves to seagrass beds to coral reefs. D.O.M. concentrations are higher in mangroves than in other systems.

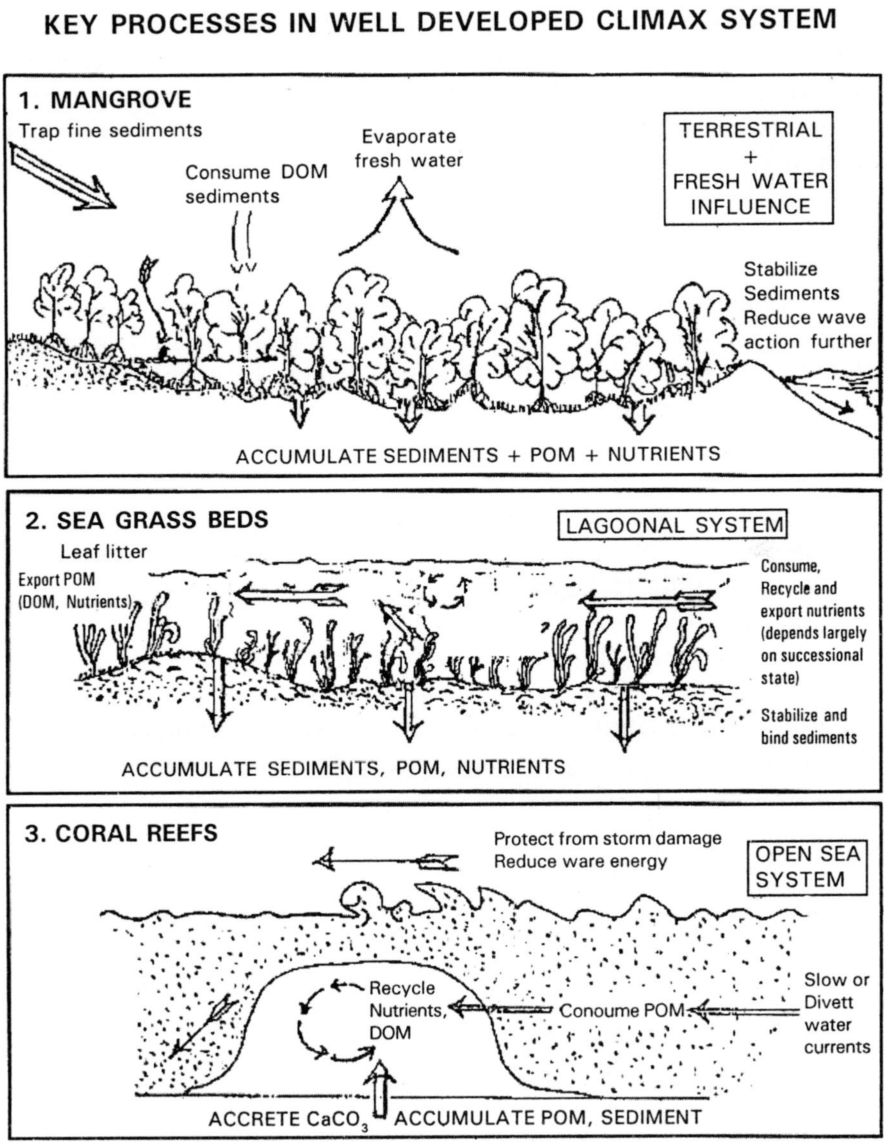

Fig. 3. Interactions between mangrove, seagrass and coral reef ecosystems

5.4 PARTICULATE ORGANIC MATTER (P.O.M.)

Particulate organic matter enters the sea in estuaries as flocculated D.O.M. and as partially decomposed mangrove and seagrass leaf detritus.

Mangroves export a large portion of their net production as leaf and woody litter. Water currents transport this material, as microorganisms, zooplankton, ciliates, nematodes and the other organisms are acting it upon. These organisms break the detritus particulates constitute a protein rich food source for larger marine organisms.

5.5 ANIMAL MIGRATIONS

Animal migrations form an important link between mangroves, seagrass beds and coral reefs. Two types of migrations are known : 1) short term migrations and 2) life history migrations between systems. The net result is a transfer of energy from the system where feeding or development occurs to the system, which shelters the adults.

Mangrove areas and seagrass beds are recognized as important nursery areas for many important commercial and forage organisms as well as some of the reef species.

The nursery function of the mangroves and seagrasses is due mainly to the availability of shelter for juvenile organisms and to an abundant supply of organic detrital food. Numerous and abundant organisms have their earlier life stages in the mangrove and seagrass regions and later move away from these systems.

After lobster spawn, the larvae settle among the roots of mangroves or in seagrass beds. As the juveniles grow and the mangrove roots or the seagrass meadows no larger provide sufficient shelter, they moveoff to the deeper surge channels and reefs where they generally hide by day and emerge at night to feed.

The snappers are more dependent on the mangrove than the grunts. The juveniles of pink shrimp live in mangrove areas and seagrass meadows, due to availability of abundant detrital food, shelter and reduced salinity.

As they grow larger they will bury in the sand by day and move into the seagrass meadows or other vegetated areas to feed at night. It is due to these juveniles and their migrations that seagrass meadow and mangrove regions are excellent fishing grounds.

Extensive epiphytic growth is known to cover mangrove roots. These organisms produce planktonic larvae; a certain part of these larvae will be carried by water movements to the other systems and serve as a potential food source there.

5.6 HUMAN IMPACT

Human impacts disturb the system varying degrees, ranging from trivial amounts to complete destruction. Disturbing factors may be physical, chemical or biological (for example tree felling, nutrient pollution or over fishing) but each can cause disturbance in their modes and thus affect any of the flows between the systems. In particular, they may affect the sediment control function, the release of dissolved and suspended nutrients or migrations of economically important animals.

In the case of mangrove, seagrass and coral reef communities, extreme disturbance means destruction, there are no alternative systems.

Chapter 6

IMPACT OF CLIMATIC CHANGE AND SEA LEVEL RISE ON MANGROVES

6.1 SCENARIOS OF GLOBAL CLIMATIC CHANGE

Climatic models suggest that the observed changes in global mean surface temperature and sea level could be the result of the increased concentration of green house gases (technically, Rediatively Important Gases, RIGs) in the atmosphere which result from anthropogenic activities including increased agricultural production, deforestation and the burning of fossil fuels.

These gases are carbon dioxide (CO_2), nitrous oxide (N_2O), methane (CH_4), tropospheric ozone (O_3), chloroflurocarbons (CFCs), carbon tetrachloride and dozens of trace gases.

Increasing concentration of these gases released by anthropogenic activities could heat the earth to temperatures warmer than at any time in the two million years and thereby accelerate the rates of sea level rise.

A planet's temperature is determined by the amount of sunlight it receives the amount of sunlight it reflects and the extent to which its atmosphere retains heat. When sunlight strikes the earth, which radiates the heat as infrared radiation. However, carbon dioxide, nitrous oxide, methane, ozone, chloroflurocarbons and a few other trace gases found naturally in the atmosphere, allow short-wave visible radiation to enter the atmosphere, but absorb and reradiate (i.e. trap) long-wave thermal radiation. The atmosphere traps heat and warms the earth in a manner somewhat analogous to the glass panels of green house, this phenomenon is generally known as the green house effect, the relevant gases are known as green house gases. Without green house effect of the gases that occur naturally, the earth would be 33 degrees C (63 degrees F) colder than it is currently (Hansen, 1981, 1984).

Since the industrial revolution, fossil fuel combustion, deforestation, land degradation and cement manufacture have released enough carbon dioxide into the atmospheric concentration of carbon dioxide by 20%, the concentration has increased 10% since 1958. Every year more than 7 billion tonnes of carbon are released into the atmosphere. About 5.6 billion tonnes of carbon are released every year due to fossil fuel combustion, which is the largest source of carbon.

Carbon cycle Modelers and energy Economists generally expect the concentration of carbon dioxide to increase 50% by 2050 and to double by 2075. Atmospheric carbon dioxide will be doubled from about 360 ppm now to over 700 ppm by 2100 (Mitchell & Gregory, 1992).

About one third methane emissions are due to agriculture, mainly rice farming and cattle rearing. Other sources are leakes from natural gas pipelines, oil wells and coal seams, burning trees and grasses, landfills, sewage treatment and termites.

Nitrous oxide is believed to originate from fossil fuel combustion, use of nitrogenous fertilizer, forest burning and change in land use.

Fossil fuel combustion is the main source of ozone of troposphere. This lower level ozone is formed from a combination of nitrogen oxides and non-methane hydrocarbons.

Chloroflurocarbons (CFCs) gases are manufactured for use in aerosol propellants, blowing agents for plastic foams, refrigerants and solvents.

Halons, compounds containing bromine and used in certain specialized fire-extinguishing systems, are closely related to chloroflurocarbons and have similar effects (Houghton, 1990).

In theory, past emissions of these gases have committed the earth to a warming of 1-2.5 degrees C over pre-industrial levels. The fact that global mean surface temperature has arisen around 0.5 degrees C (0.6 ± 0.2 degrees C) over the last century is now widely accepted (Houghton, 1991).

Under the Intergovernmental Panel on Climatic Change (IPCC) Business-as-usual scenario for future green house gas emissions global mean surface (air) temperatures are projected to rise by around 0.3 degrees C per decade, a rate of increase greater than that seen over the past 10,000 years (Houghton, 1991). The majority view of scientists is that the global mean temperatures will have rise to 1.8 degrees C above pre-industrial levels (range 1.3-2.5 degrees C) with an increase in global mean precipitation and evaporation of 3%. This rate would lead to a likely increase in global mean temperature 2.4-5.1 degrees C (best estimate 3.5 degrees C) and precipitation 7% greater (IPCC, 1990 and Trexler *et al*, 1990).

Wigley and Raper predictions (1992) using Intergovernmental Panel on Climate Change (IPCC) 1992 Scenario A (IS92a), one of six alternative scenarios offered by IPCC in 1992 suggest a best guess global mean warming of 2.5 degrees C (range 1.7 to 3.8 degrees C) by the year 2100. This corresponds to a 0.2 degrees C warming per decade.

One positive effect of increased carbon dioxide is that global net primary production may increase by 20% to 26% for a doubling of CO_2 (Mellillo *et al*, 1993).

The increase in atmospheric CO_2 has the potential to increase the growth of C_3 plants by stimulation of photosynthesis, depressing respiration, and relieving nutrient and high temperature stresses (the carbon dioxide fertilization effect) (Mellillo *et al*, 1990).

The most obvious result of such warming is a rise in sea level due to thermal expansion of the oceans and melting of some land ice.

The sea level at any shoreline varies second by second as well as year by year to give an extremely complex signal in which long term systematic sea level changes are difficult to separate from short term cyclic changes caused by waves and tides. Long term sea level changes may be caused by two factors. The first factor is caused by an increase in the temperature of the ocean or the melting of fresh water stress in ice caps. As a result changes in the volume of water held in the World Ocean. These changes are known as eustatic changes. These are the sea level increases envisaged as occurring due to global warming in which an increase in carbon dioxide and other gases traps the increasing solar radiation and increases earth surface temperatures. The second factor may be due to a number of causes. These changes occur in the level of land relative to the sea and are known as isostatic change, which results in an apparent rise in sea level.

The sea level is currently rising at 1.6-3.3 cm per decade, one quarter of which is due to the thermal expansion of the oceans and the rest to the melting of glaciers and snow fields. Between now and 2030 it is predicted to rise by about 20 cm (range 10-32 cm) and by 2070 by 45 cm (range 33-75 cm). Some have put forward higher estimates, 40-120 cm (Trexler *et al*, 1990), while most pessimistic scenarios predict rises of several meters. Wigley and Raper (1992) have revised the best guess projections of sea level rise. Their predictions using IS92 suggest a best guess global mean sea level rise of 48 cm (range 15 to 90 cm) by the year 2100. This corresponds to a rise in sea level of 4 mm/year. The worst outcome would be melting of the west Antarctic ice sheet, which would cause a rise of 7 meters in the sea level. Even the more modest predicted rise would totally submerge some Island groups, erode barrier Islands, and inundate many coastal wetlands and estuaries. There would be saline intrusion into ground water supplies. Low lying coastal hinterlands, which include many of the Worlds's most productive and densely settled areas would be lost. The potential impact of a sea level rise of 1 meter has been described as follows :

"Worldwide, the land area that would be subject to inundation or made vulnerable to salt water intrusion is about 500 million hectares. This is only about 3% of all land area, but it constitutes over 30% of the most productive cropland area. As many as one billion people now live in the vulnerable areas, including some very large cities ... as much as one fifth of World market valued assets could be adversely affected" (Walter and Ayres, 1990).

Over the past 100 years global mean temperature has arisen by about 0.45 ± 0.15 degrees C and sea level by about 10 to 15 cm (Folland *et al*, 1990; Warrick & Oerlemans, 1990).

Thus rates of warming are predicted to be about five times and rates of sea level rise about four times faster those that have occurred over the century.

Studies of past climatic fluctuations and global circulation models (GCMs) suggest that warming in tropical regions will be less than the global mean warming (Folland *et al*, 1990 and Mitchell *et al*, 1990). Therefore, the changes in temperature will be greater in higher latitudes than at the equator. Thus North America, the USSR, Europe and China will be more affected than tropical region. Holdgate *et al* (1989) suggest that the tropics will warm by 0.7 to 0.9 times the global average, which would indicate a best guess warming for the tropics of perhaps 2 degrees C by 2100. Palaeoclimatic studies of the warm periods of the Holocene optimum (5000 to 6000 years B.P.) and the Eemain interglacial (125,000 to 130,000 years B.P.) indicate that despite average summer temperature in high northern latitudes 3 to 4 and 4 to 8 degrees C respectively warmer than today, there was no detectable warming in the tropics (Folland *et al*, 1990).

Use of palaeoendemics as analogs of future warming must be treated with caution because of the different factors involved - changes in seasonal distribution of solar radiation vs. globally averaged rise in Green house gas concentrations.

IPCC process evaluating scientific understanding, impacts and response measures is on going and hence it is to be expected that these estimates will change in any future assessment. It is to be noted that these scenarios refer to changes in global mean values and that the actual values of temperature and sea level at any particular area may differ considerably from these estimates due in part to site specific orographic climate effects, geomorphological processes and local sources of anthropogenic change (Pernetta, 1993).

6.2 LARGE SCALE IMPACT ON MANGROVE BIOME

A major constraint to impact assessment in mangroves is the absence of accurate scenarios of future rainfall, both seasonality and total quantity. Not only is rainfall a major determinant factor in limiting the distribution of terrestrial biomes on a global scale it has major effects on a local scale though its influence on processes such as water and sediment flux from land to near shore waters. Without some detailed assessment of such changes on a local and subregional scale precise assessment at single location is not possible (Pernetta, 1993).

Mangroves are characterized by zonal organization of constituent species, which has often been interpreted as representing successional change over time (Davis, 1940; Chapman, 1976) by high rates of organic production and by accumulation of vegetative detritus *in situ* to form mangrove peat (Ellison & Stoddart, 1991). This mangrove peat also contain faunal remains and allochthonous inorganic and organic matter brought in by the tides and by rivers, though the magnitude of these components depend on the tidal range and physiographic location. The accumulation of peat under mangroves in the upper half of the tidal range renders it a useful sea level indicator (Scholl, 1963, 1964).

So peat build up in mangrove communities is an important mechanism for progression of the mangrove ecosystem seaward. In the Sunderbans area of the Ganges and Brahmaputra deltas in Bangladesh, the Florida Everglades, the deltas and estuaries of northern and northeastern Australia, the Niger delta and the coasts of Papua New Guinea, Borneo and Sumatra where terrigenous inputs are more, some mangroves, have prograded seaward at rates of tens of centimeters or more per year, by trapping sediments and mangrove tree litter in their root systems. On low lying coral Island shores of Maldives, Grand Caymen and Tongatapu and on arid coastlines or Red Sea and Arabian Gulf where terrigenous inputs of sediments are lacking, rates of accumulation of sediment and litter are much slower (Ellison and Stoddart, 1991).

Therefore, the amount of allochthonous sediment available will be a major factor in determining the persistence of mangroves during relative sea level rise.

6.3 IMPACT OF SEA LEVEL RISE ON MANGROVES

During the early Holocene when sea level was rising upto 2 m/100 years, large mangrove swamps did not exist. As sea level stabilized at various dates in different regions around 6500 BP than the first extensive mangrove ecosystems established in sheltered locations (Ellison & Stoddart, 1991).

Studies of mangrove peat cores from low flat Islands without significant inputs of terrigenous sediments indicate vertical accretion rates of only 8-9 cm/100 years with some indications that between 7000 and 5500 years BP such mangrove systems have kept up with a sea level rising at 12 cm/100 years (Ellison and Stoddart, 1991). These mangroves could thus become under stress at current rates of eustatic sea level rise (Snedaker *et al*, 1994). Studies from high Islands with significant terrigenous input of sediments indicate vertical accretion rates of up to about 22 cm/100 years (Bloom, 1970 and Ward, 1988).

Thus neither low flat nor high Island mangroves appear to be able to accrete vertically as fast as the best guess rate of sea level rise of 40 cm/100 years. Also as sea level rises, so will the shorelines erosion tend to increase (Bruun, 1962) with increased removal of litter, detritus and sediment from the mangrove edge by high tidal flushing. Mangrove stands in such situations will thus tend to be compressed and may suffer reductions in species diversity.

It is to be noted that mangroves do not create accreting shorelines (Percival & Womersley, 1975). The presence of mangroves in accreting shorelines led to the mistaken belief that mangroves were responsible for causing sedimentation and hence creating land. Once established in such situations, mangroves may hasten the sediment deposition through reducing the wave action and by forming sediment traps with their complex, inter twined root systems. In Australia (Western Port and Cairns Bay), Bird and Rosengren (1985) have shown that *Avicennia marina* growing on a muddy substrate acted as a sediment trap where wave action was weak and fine grained sediment available. As a result the shoreline was progressed and builds up of a depositional terrace in the upper intertidal. But such mangrove tree species are likely to be weaker in the face of accelerated sea level rise than others. Vicente *et al* (1989) have noted that in the Caribbean, the aerial roots of *Avicennia germinans* project only short distances above intertidal muds, while prop roots of *Rhizophora mangle* stand much higher above mean sea level (sometimes more than 6 m).

Therefore, suggested that in the relative short term rising sea level may cause massive local extinctions of *Avicennia* but have insignificant effects on *Rhizophora*.

Studies on mangrove peat accumulation from Tongatapu of Tonga, a low flat limestone Island, found average peat accumulation rates (during a period of falling sea level when rates are higher) of only 26 cm/100 years under higher swamp species of the genera like *Bruguiera* and *Excoecaria*, while the rates are as high as 53 cm/100 years beneath the lagoon-fringing *Rhizophora* (Ellison and Stoddart, 1991).

The amount of allochthonous sediment available will be a major factor in determining the persistence of mangroves during relative sea level rise. Ellison & Stoddart (1991) stated that in areas where allochthonous sediment source are small and absent mangrove systems may undergo catastrophic collapse under scenarios of accelerated sea level rise. It should also be noted that massive allochthonous sediment being washed to the coast as a result of poor agricultural practice can bury pneumatophores and kill mangroves (Stoddart & Pethick, 1984).

Pernetta and Osborne (1990) suggested that sea level rise may reduce sediment supply and lead to a recession of the coastline, compression of mangrove zonation and reduction in its extent by over 60%.

In Andaman & Nicobar Islands saline soil with alluvial formation or soft muddy substratum below, irregular and deeply indented coastline which giving rise to a number of tidal creeks dominate the mangrove system. But no information existed concerning the estimation of river discharge of water and sediment. However, researches on mangrove soils in Andaman-Nicobar Islands have been limited to few localities. The result of which shows that soils upon which mangroves develop consist principally of sand (upto 75%) (Singh *et al*, 1988).

Studies on physical and chemical properties of mangrove soils show that the bulk densities of the surface soils slightly decreased from the open area to the mangrove forests on the landward regions and sand content increased but the percentage of silt and clay diminished in the more Inland zones (Dagar *et al*, 1991). Thus moisture retention capacity of the soils is rather low. Infiltration rate is moderate and depending upon the soil and nature of soil cover runoff rates may be medium to very high. These uncovered soils are subjected to serious soil erosion.

However, a simplistic first approximation of the impacts of a given rise in sea level can be made merely by shifting the shoreline an appropriate distance Inland and replotting mangrove zonation in relation to the new conditions of elevation.

Under conditions of rising sea level, areas of mangrove growing in deltic and tidal coastal plains are likely to undergo erosion and loss from the seaward edge. Simultaneously, as saline conditions penetrate further Inland, the mangroves of the inner margins will begin to replace the terrestrial and freshwater swamp vegetation. Although, the rate of landward extension may not be the same as the rate of loss of pioneer mangroves from the seaward edge of the system (Pernetta, 1933). If the rainfall, erosion rates, current rates of crustal downwraping and sediment compression remain the same, then under conditions of rising sea level the rate of progression of mangrove zone may be expected to decrease.

Accordingly, supplies of sand by long short drift will be diminished and regression of the mangrove system in the Islands may be expected to accelerate.

As noted above, *Rhizophora* communities likely to be more robust to rising sea level over a time scale of many generations having found peat accumulation rates beneath *Rhizophora* to be double those beneath the special of *Bruguiera* and *Excoecaria*. Moreover, the prop roots of *Rhizophora* species stand much higher above mean sea level.

In Andaman-Nicobar Islands, the species *Rhizophora apiculata* and *R. mucronata* are the major representative of the proximal zone where the mangrove formation has extended into the sea so that there is no foreshore and the community is inundated for most of time.

Therefore, it is to be expected that there will be no significant effects on *Rhizophora* under future scenarios of sea level rise.

The viable longevity of the propagules is 12 months for *Rhizophora*, whereas it is 110 days for *Avicennia*.

The *Avicennia*, has the smallest propagule, has the highest rate of seedling mortality, while the *Rhizophora*, with the largest propagule has the lowest seedling mortality rate (Rabinowitz, 1978).

The relative changes to the major mangrove zones are unlikely to be equal regression of the seaward *Sonneratia*, *Avicennia* zone is likely to occur at a faster rate than landward migration of the *Rhizophora* and *Bruguiera* zones. Hence differential loss or reduction in the abundance of species in the pioneer mangrove zone is likely to occur (Pernetta, 1993).

In the Gulf of Papua, propagules of pioneer species such as *Avicennia eucalyptifolia* and *Sonneratia lanceolata* appear capable of withstanding longer periods of inundation than those of other species (Floyd, 1977). The vertical pneumatophores of the species *Avicennia* and *Sonneratia* modify local currents and promote further sedimentation (Spercely, 1977), such that once the level of the substratum has arisen, propagules of *Rhizophora* species and later *Bruguiera* become established.

The productivity of commercial and subsistence fisheries occurring through the mangrove system and the zonation and productivity of the entire system will change dramatically under the changes in the overall extent mangrove ecosystems combined with changes the relative extent of different zones.

6.4 IMPACT OF TEMPERATURE ON MANGROVES

A mean sea-surface and air temperature rises of 2°C by 2100 are likely to be beneficial to mangrove species living near the poleward limits of present day distributions leading to increased species diversity, greater litter production and larger trees in these peripheral mangrove systems. There is also potential for limited poleward range extensions of mangroves (Edwards, 1995). Extension of range into areas, which become climatically suitable in a warmer World, will involve the changes in species growth and reproduction. As conditions at a particular site approach more closely the optimum conditions presently occurring towards the centre of the biomes present distribution, individual species vigour and reductive rates can be expected to increase. Thus, increased species diversity at the community level will add to the competitive ability of mangrove communities as a whole. Outside the present latitudinal limits for mangroves, comparable sedimentary coastal environments are generally occupied by salt marsh vegetation. It is likely that a gradual replacement of salt marsh vegetation by scrubby mangrove, first of *Avicennia* and later of *Rhizophora* may be expected to occur (Pernetta, 1993).

The possibility of thermal stress affecting mangrove root structures and establishment of mangrove seedlings at sea temperature above 35°C has also been noted (Anonymous, 1992).

6.5 IMPACT OF CARBON DIOXIDE

Most obligate mangroves are C_3 plants. They have a C_3 pathway for a carbon fixation during photosynthesis (Clough *et al*, 1982). Thus should also benefit significantly from increased atmospheric CO_2 concentrations with increased productivity. Despite living in wet conditions mangroves have remarkably low transpiration rates. This adaptation may be linked to problems of salinization of the soil around the roots of the mangroves as water is taken up because of the exclusion of most salt at the root surface (Passioura *et al*, 1992). Not only does the rate of photosynthesis increase but reduced stomatal conductance under higher CO_2 level leads to more efficient water control through reduced transpiration losses. As noted above, one consequence of increased CO_2 may be an increase in overall primary production in mangroves and hence an overall increase in litter production and vertical accretion rates.

These amending circumstances are likely to be little importance compared to the impact of accelerated sea level rise.

Litter production in Andaman & Nicobar Islands has been estimated for the genus *Excoecaria* 1.1 kg m^{-2} yr^{-1}, for *Bruguiera* 1.0 kg m^{-2} yr^{-1}, for *Rhizophora* 0.95 kg m^{-2} yr^{-1} and *Avicennia* 0.8 kg m^{-2} yr^{-1} (Dagar & Sharma, 1989). In mud flats and creeks of Phoenix Bay and Corbyn's Cove of South Andaman a figure of 2.0-3.6 gc/m^2/day appeared to be highest value of primary productivity recorded to date (Ramachandran Nair & Gopinathan, 1983).

6.6 IMPACT OF ULTRAVIOLET RADIATION ON MANGROVES

Depletion of tropospheric ozone layer caused largely by chloroflurocarbon compounds (notably CFC-11 and CFC-12) increases surface exposure to ultraviolet radiation with effects being most marked at high latitudes (Isaksen, 1989).

It is predicted that globally averaged total column ozone may decrease by about 1 to 4% by 2035 (Eckman & Pyle, 1989) and average surface ultraviolet radiation exposures in the tropics may be expected to increase by about 2 to 8% above today's levels (Edwards, 1995). Predicted changes in surface ultraviolet radiation levels are not really climatic changes but are global in impact.

Mangroves have ultraviolet absorbing capacity due to the presence of some phenolic compounds in their leaves. There is also some evidence that ultraviolet radiation may stimulate the production of these phenolic compounds in some species of *Bruguiera* and *Rhizophora* (Lovelock *et al*, 1992).

Thus mangroves seem to have the potential to adapt to the increase in incident ultraviolet radiation.

Compact data on impacts of global climate change are largely lacking with important unanswered questions from the molecular biochemical level to the ecosystem level. There are also considerable uncertainties in the predictions of just how climate will change and how fast and by how much sea temperature in the tropics may rise (Edwards, 1995).

However, on the basis of above discussion it might be noted that the response of individual species to single source of environmental change may be either positive or negative. The combined effect of environmental changes to several variables may be either positive, neutral or negative depending on the magnitude of the changes to individual species response (Pernetta, 1993). A general review of the potential responses of mangrove species and ecosystems to changes in environmental factors as a consequence of global climatic change and sea level rise was provided by Pernetta (1993).

Mangroves may be under threat from global climatic change and accelerated sea level rise over the next century but man is the major threat at present. So, any discussion of the potential impacts of future climate change and sea level rise on a mangrove ecosystem would be pointless if it were conducted in isolation from a consideration of the anthropogenic factor currently affecting the status of the mangrove ecosystem. This anthropogenic impact may be for more important locally than global climate change.

World wide tropical coastlines are currently subject to increasing stress as a consequence of anthropogenic activities in combination with increasing human population density (Pernetta and Elder, 1992). Around two thirds of the World's population currently lives in close proximity to the coastline. As a consequence coastal wetlands, including mangroves occur on some one quarter of tropical coastlines are coming under increasing pressure from clear felling (for firewood, woodchips, charcoal, pulp and rayon), coastal agriculture (rice and coconut production), mariculture (e.g. for prawns), reclamation for housing and industrial developments, pollution (from both domestic and industrial sources) and oil spills.

One further source of stress of mangrove systems is accelerated subsidence of delta areas due to ground water extraction (Milliman, 1989) and damming of rivers upstream drastically reducing water, sediment and nutrient supply.

Thus, the mangrove ecosystems are already stressed by man's activities, facing the prospect of further stress from climate change and sea level rise. These more localized anthropogenic impacts may also seriously impair the abilities of the ecosystems to respond to the stresses and demands of global climatic change.

Significance of mangroves in the lives of human populations, their function in coastal sediment stabilization, as well as their intrinsic scientific importance as ecosystems and recorders of biotic diversity indicate that particular attention be paid to their survival.

Chapter 7

MANAGEMENT IMPLICATION

The conservation and management of mangroves has to be considered as a special case, different from most other ecosystems. This is mainly because mangroves represent interface ecosystems that straddle the land and the sea, from freshwater to seawater, often with distinct zones of plant and animal species and are closely linked not only with atmosphere above (like most other ecosystems) but also with the adjacent landward catchment and seaward marine system. Thus conservation of mangroves should include not only the various vegetation and tidal inundation zones but also the adjacent marine and terrestrial areas (including water catchment areas). So, it is unrealistic to just conserve a patch of mangrove as we would a patch of terrestrial rain forest.

7.1 CONSERVATION

The term conservation was originally derived from the phrase "Conservation of wise use" (Darling, 1964). The Oxford Concise Dictionary (6th Edition, 1976) defines conservation as "preservation especially of natural environment" and conserve as "keep from harm, decay or loss, especially with view to later use".

It becomes apparent that depending on the situation, there may be differences in degree of preservation, from a total "hands off" situation to one of "wise use", "sustainable use", "later use" and a host of other more liberal uses.

To conserve mangrove ecosystem in Andaman & Nicobar Islands it is necessary to adopt a pragmatic rather than idealistic approach. The two broad categories of conservation should have to go hand in hand. The first category is one of conservation for posterity and the second for long term prosperity. In other words, there should have some areas of representative mangroves legislated a national park or natural heritage areas (i.e. conserved for posterity) and most of the rest of mangroves conserved through sustained yield management (i.e. for prosperity).

In Andaman & Nicobar Islands, some percentage of the total area of mangroves should be conserved as National Parks. At the same time this will ensure that the areas set aside are large enough and adequately buffered to survive in their natural undisturbed state.

Since mangroves are non-homogeneous open ecosystems and are extremely dynamic; interacting with the atmosphere above as well as with processes of the adjacent land and sea, a whole strip of land from the freshwater to the seawater will be required. Thus the boundary of the mangrove national park should extend into the sea as well as into its freshwater catchment so as to allow space for natural growth.

7.2 MAJOR THREATS TO MANGROVES

Mangrove ecosystems of Andaman & Nicobar Islands are being threatened by both natural and man induced stressors. Both of these two stressors can impinge on the system in a chronic or acute mode. For example, hypersalinity can be a chronic natural stressors in many areas. Storms and droughts are examples of acute events. In polluted areas some pollutants can be chronic stressors, or behave as acute stressors during a spill. Strategically, to effectively protect or conserve our mangrove resources, it would be wise to identify and understand the main threats to these resources. Some of the natural and man made threats to the mangrove ecosytem of Andaman & Nicobar Islands are as follows:

Population Pressures :

Burgeoning populations are possibly the biggest cause of mangrove destruction and degradation. Increased population pressures and over exploitation over the last 100 years has led to noteworthy reduction in the floristic components so much so, only the very robust species are encountered in the secondary formation. Mangrove areas are rich sources of firewood, charcoal, housing materials and fishing gear, tannin, food and drink, medicine, aquatic products such as various kinds of fish and shrimps, molluses, crabs and other animals. Therefore, mangrove areas are a primary source of livelihood for a large part of World's population. Mangrove areas also produce biological resources and sustain functions that are crucially important to the local, regional and global environment.

However, economic development, rapid population growth and migration from Mainland to Island mangrove areas exert increasing pressure on mangrove resources. Human settlement in mangrove areas is another but more serious exploitation of these resources. Since 1911, the population of Andaman & Nicobar Islands was increased more than seven folds till 1991 Census (Fig. 4).

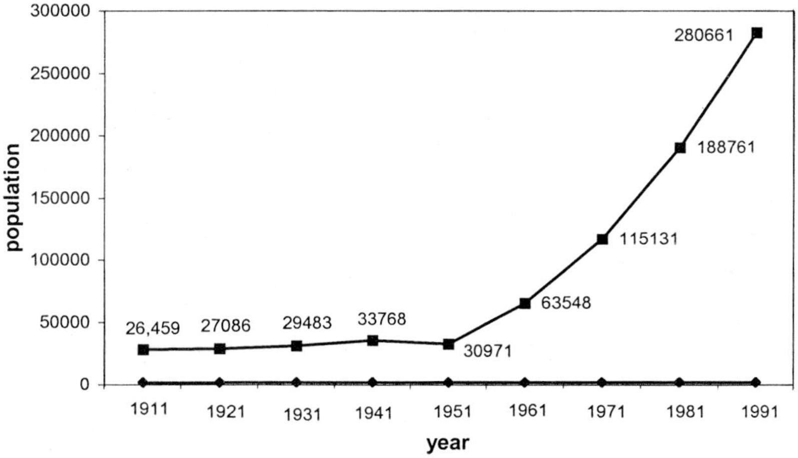

Fig. 4. Graph showing population of Andaman & Nicobar Islands since 1911 to 1991

Like all systems, mangroves and coastal areas have a certain carrying capacity. Once this capacity is exceeded, the system will be stressed and will start to break down. The increased human population results in clearance of large tracts of virgin mangrove forests for conversion into agriculture land, house construction, domestic purposes, etc. and to fulfil the basic requirements of human needs. Thus all we can do is hope that the rate of population growth can be brought down fast enough before irreversible damage to the environment resuslts.

This population problem is the basis of all the other following threats.

Land Reclamation :

This is specially related to population pressure as well as less than well-informed or unenlightened authorities "reclaiming" mangrove lands for housing estates. It is a quick way to make a fast buck where demand for land is high (Jin-Eong, 1995).

Illegal Deforestation :

It is common practice by the local people in the mangrove areas, reserved forest areas and in the unclassified forest ranges. The local people to meet the local demand are also gradually removing a fewer mangrove patches in the reclaimed areas. There was a large patch of *Nypa fruticans*, commonly known as Dhani, along the both side of the 'Dhaninala' of Middle Andaman and 'Dhanikhari' of South Andaman. But now it is totally removed from the said areas due to illegal deforestation.

Oil Pollution :

Petroleum and its byproducts injure and kill mangroves in a variety of ways. Crude oil coats roots, rhizomes and pneumatophores and disrupts oxygen transport to underground roots. Petroleum is readily absorved by lipophylic substances on surfaces of mangroves. This leads to severe metabolic alterations such as displacement of fatty mollecules by oil hydrocarbons leading to destruction of cellular permeability and/or dissolution of hydrocarbons in lipid components of chloroplasts (Baker, 1971). As with other intertidal communities, many of the invertebrates, fishes and plants associated with the mangrove community are highly susceptible to petroleum products.

The western part of the Indian Exclusive Economic Zone, i.e. Arabian Sea adjoining peninsular India, forms the main international tanker route for oil tankers originating from the Gulf. It has been estimated that some 434 million tonnes of crude oil is transported annually along this route, involving approximately 2500 laden tankers. The preferred route is through 9° channel between Maldives and Lakshadweep Islands, during the southwest monsoon (May to September) and north of Lakshadweep following the 200 metre depth curve west of Mangalore, at other times. Considering the large volume of oil transported and high rate of tanker movement the probability of tanker accident is high. The last major accident is the area occurred in January 1993 and a few tonnes of oil spilled in the Andaman Sea. During the southwest monsoon, oil spills occurring between 68°E and 76°E, will tend to drift towards the Lakshadweep Islands and Kerala Coast. At other times oil spills could threaten the Lakshadweep between December and February and the Kerala coast during March and April. Any accidental spilage oil along the tanker route will cause severe and in some cases irreparable damage to the marine ecosystem. Similar damage would be caused if accidents take place during southwest monsoon seasons. Besides such oil transfer facilities in India, the lightening operations carried out in the rest of the South Asian countries may also be the additional sources of oil pollution to the Indian waters especially during accidents.

Sedimentation :

Mangroves are adapted to high sedimentation environments, but sudden deposition of large quantities of sediments can cause mortality. Under natural conditions, excessive sedimentation occurs as a result of catastrophic phenomena like storm generated waves or floods. Man, however, is often cause the severe sedimentation problems in some mangrove areas.

Deforestation and loss of soil can increase the rate of sedimentation in low lying areas, causing shallow lagoons to become salt flats and killing inner swamp trees.

Natural Factors :

Coastal retrogradation processes and cyclones are two significant natural factors which destroying mangrove vegetation in the Andaman & Nicobar Islands.

With retrogradation process of coastline, mangrove vegetations are being degraded in the Andaman & Nicobar Islands. Channel bank erosion in the concave beds, point-bar migration in meandering tidal channels and rotational slumps of mud banks are also destroying mangrove vegetation. Mangrove trees are uprooted on the retrograding coastlines of Dugong Creek (Little Andaman), Kalighat Creek (North Andaman) and Wrafter's Creek (Middle Andaman).

Cyclonic disturbances develop in the Andaman-Nicobar Islands during the month of June through September. Mangroves are vulnerable to these disturbances due to their coastal location, their shallow root system, the poor cohesiveness or load bearing ability of most mangrove soils and their exposure to waves, surge and erosion by strong flows.

Management Practices :

Mangroves are among the most productive coastal ecosystems known to science. At no cost to man, mangrove forests provide habitat for valuable birds, mammals, amphibians, reptiles, fishes, and invertebrates and protect endangered species, at least partially support extensive coastal food webs. So, in addition to their aesthetic and recreational values, they form an important part of the economic resource base of Andaman-Nicobar Islands because of their intimate ties to coastal fisheries, seagrass beds and coral reefs.

Society derives direct benefits from the conservation of these coastal ecosystems. They are part of the resource base upon which islands depend. There are very practical reasons for their conservation, including the products they yield and their role in protecting the coasts from storm damage, as well as their more intangible values.

Mangrove ecosystms are in jeopardy in many areas of the Andaman & Nicobar Islands, due to high populations, mounting pressure to develop a limited land base and inadequate impact assessment procedures. In the small Islands, Inland and Coastal ecosystems are intimately linked. Inland activities like deforestation, road building, agriculture and many others can increase sediment levels in estuarine and coastal waters, contributing to the degradation of coastal ecosystem.

In many areas of the World mangrove swamps have been converted to other uses such as aquaculture and agriculture. Conversion to aquaculture and agriculture is cursed with a variety of problems including subsequent land subsidence and "cat clay" problem. The latter refers to the drastically lowered soil pH, which often occurs after drainage and has been traced to oxidation of reduced sulphur dioxide (Moorman and Pons, 1975). Experience in Africa, Puerto Rico and South East Asia proves that mangrove forests in their natural state are more valuable than the "reclaimed" land.

Protection of Andaman-Nicobar mangrove includes 1) preservation of the mangrove ecosystem in the natural state, 2) utilisation of mangroves on a sustainable basis, 3) prevention of conversion of mangroves into aquaculture and agriculture, 4) prevention of outright destruction from dredging and filling, 5) prevention of any alteration of hydrological circulation patterns, 6) prevention of oil spills, 7) prevention of increased wave action or current velocities from boat waves and seawalls. Where mangroves have been destroyed they can be replaced or suitable alternate areas can be planted, acre for acre, through mitigation procedures.

Important considerations (Lewis, 1979; Teas, 1979) in transplanting mangroves are 1) to plant in the intertidal zone and avoid planting at too high or too low an elevation, 2) to avoid planting where the shoreline energy is too great, 3) to avoid human vandalism and 4) to avoid accumulations of dead sea grass and other wrack.

The best management practice for all types of Andaman & Nicobar mangrove ecosystems is preservation. Central to this concept is the preservation of adjacent ecosystems that are linked significantly by functional processes. In situations where overwhelming economic pressures dictate mangrove destruction, every effort should be made to ameliorate any losses either through mitigation or through modified development.

REFERENCES

Anonymous. 1992. UNEP-10C-WMO-IUCN Meeting of Experts on a Long-Term Global Monitoring System of Coastal and Near-shore phenomena related to climate change, Pilot projects of Mangroves and Coral Reefs, UNEP-10C-WMO-IUCN/GCNSMS-113 (UNESCO).

Atkinson, M.R., Findlay, G.P., Hope, A.B., Pitman, M.G., Saddler, H.D.W., and West, K.R. 1967. Salt regulation in the mangrove *Rhizophora mucronata* Lam., *Aegialitis annulata* R. Br. *Aust. J. Biol. Sci.* 2,**7**: 547-51.

Balachandra, L. 1988. A comprehensive account of the mangrove vegetation of Andaman & Nicobar Islands. *Indian For.* **114**: 741-751.

Bird, E.C.F. and Rosengren, N.J. 1985. Mangrove and Coastal morphology pp.17-28 in Cragg, S. and Polunin, N. (eds.) Workshop on mangrove ecosystem dynamics. UNDP/UNESCO. Regional Project RAS/79/002.

————— 1977. Outlines of ecology, botany and forestry of the mangals of the Indian subcontinent. Chap. 12 in Ecosystems of the World vol.1. Wet Coastal ecosystems, ed. V.J. Chapman. Amsterdam: Elsevier Scientific.

Bloom, A.L. 1970. Paludal stratigraphy of Truk, Ponape and Kusaie, East Caroline Islands. *Geol. Soc. Am. Bull.* **81**: 1895-1904.

Bowman, H.H.M. 1917. Ecology and physiology of the red mangrove. *Proc. Amer. Philos. Soc.* **56**: 589-672.

Bruun, P. 1962. Sea-level rise as a cause of shore erosion. J. Water. Harbors Div. *Proc. Am. Soc. Civ. Eng.* **88**: 117-130.

Burkholder, P.R., Burkholder, L.M. and Almodovar, L.R. 1967. Carbon assimilation of marine flagellate blooms in neritic waters of southern Puerto Rico. *Bull. Mar. Sci.* **17**(1): 1-15.

Chapman, V.J. 1970. Mangrove phytosociology. *Trop. Ecol.* **11**: 1-19.

——————— 1975. Mangrove Biogeography, vol.1. pp.3-22. In: Walsh, G.E.S.C. Snedaker and H.J. Teas. *Proceedings of the International Symposium on Biology and Management of Mangroves*, Univ. Florida Gainesville.

——————— 1976. Mangrove Vegetation. Vaduz : Cramer, 447 p.

——————— 1977. Introduction. pp.1-29. In: *Ecosystems of the World* 1. Wet Coastal ecosystems. Elsevier, Amsterdam.

Chopra, R.N., Nayar, S.L. and Chopra, I.C. 1956. Glossary of Indian Medicinal Plants. CSIR, New Delhi.

Cintron, G., Lugo, A.E., Pool, D.J. and Morris, G. 1978. Mangroves of arid environments in Puerto Rico and Adjacent Islands. *Biotropica*, **10**: 110-121.

Clough, B.F., Andrews, T.J. and Cowan, I.R. 1982. Primary productivity in Mangroves. In: Clough, B.F. (ed.) Mangroves Ecosystems in Australia - structure, function and management. AIMS with ANU Press, Canberra, Australia.

Dagar, H.S. 1989. Plant folk medicines among Nicobarese tribal of Car Nicobar Island. *India Econ. Bot.* **43**(2): 215-224.

Dagar, J.C. and Dagar, H.S. 1986. Mangroves and some coastal plants in ethnobotany of the tribals of Andaman and Nicobar Islands. *J. Andaman Sci. Assoc.* **2**(2): 33-36.

——————— and Sharma, A.K. 1989. Impact of developmental activities on litterfall and mangal formation in the Andamans. Paper presented in Symposium Development *vis-à-vis* Ecology. INTACH-Forest Dept., Port Blair, 1989.

———————, Mongia, A.D. and Bandyopadhyay, A.K. 1991. Mangroves of Andaman Islands. Oxford and IBH Publishing Co. Pvt. Ltd., New Delhi, p.19.

Darling, F. 1964. Conservation and ecological theory. *J. Ecology* **52**: 39-46.

Davis, J.H. 1940. The ecology and geological role of mangroves in Florida. *Paper Tortugas Laboratory*, **32**: 303-412.

Dawes, C.J. 1981. Marine Biology. John Wiley & Sons, New York.

Dixon, John A. 1989. "Valuation of mangroves", Tropical Coastal Area Management 4(3), Manila, December.

Dorairaj, K. and Soundararajan, R. 1987a. Coastal environment and fishery resources of the Bay Islands. In: *Proc. Symp. on Management of Coastal Ecosystems and Oceanic Resources of the Andamans.* Andaman Sci. Assoc., Port Blair. pp.40-44.

———— and ———— 1987b. Preliminary studies on brackish water fish and prawn farming in Andamans. *Ibid.* pp.80-87.

Eckman, R.S. and Pyle, J.A. 1989. Numerical modelling of ozone perturbations, in Ozone Depletion : Health and Environmental consequences, Johns, R.R. and Wigley, T., Eds. (Chichester: John Wiley & Sons, 1989), pp.27-42.

Edwards, Alasdair, J. 1995. Impact of climatic change on Coral Reefs, Mangroves, and Tropical Seagrass Ecosystems. In: Eisma, Doeka (ed.) *Climate change Impact on Coastal Habitation.* Florida, Lewis Publishers. pp.224.

Egler, F.E. 1948. The dispersal and establishment of red mangroves, *Rhizophora* in Florida. *Caribb. For.* **9**(4): 299-319.

Ellison, J.C. and Stoddart, D.R. 1991. Mangrove Ecosystem collapse during predicted sea level rise : Holocene analogues and implications. *J. Coastal Res.* **7**(1): 151-165.

Farber, Stephen. 1987. The value of coastal wetlands for protection of property against hurricane wind damage. *Journal of Environmental Economics and Management* **14**(2), June.

FAO 1982. Management and Utilization of Mangroves in Asia and the Pacific. Rome.

Floyd, A.G. 1977. Ecology of the tidal forests in the Kikori-Romilly sound area Gulf of Papua. *Ecology Report No.459 pp. P.N.G.* Department of Primary Industry, Division of Botany, Lae, Papua New Guinea.

Folland, C.K., Karl, T. and Vinni-Kov, K.Y. 1990. Observed climate variations and change in climate. *The IPCC Scientific Assessment*, Houghton, J.T., Jenkins, G.J. and Ephraums, J.J., Eds. (Cambridge: Cambridge University Press, 1990), pp.195-238.

Foo, H.T. and Wong, J.T.S. 1980. Mangrove swamp and fisheries in Sabah. *Trop. Ecol. Dev.* **2**: 1157-1161.

Gill, A.M. and Tomilson, P.B. 1969. Studies on the growth of red mangrove (*Rhizophora mangle* L.) I, Habit and general morphology. *Biotropica* **1**(1): 1-9.

Hamilton, L.S. and Snedaker, S.C. (Eds.) 1984. Handbook for mangrove area management. United Nations Environment Programme and Environment and Policy Institute, East-West Centre, Honolulu, Hawaii.

Hansen, J.E., Johnson, D., Lacis, A., Lebedeff, S., Rind, D. and Russell, G. 1981. Climate Impact of Increasing Atmospheric Carbon Dioxide. *Science* **213**: 957-966.

Hansen, J.E., Lacis, A.A., Rind, D.H. and Russell, G.L. 1984. Climate Sensitivity to Increasing Greenhouse gases. In: Barth, M.C. and Titus, J.G. (eds.) *Greenhouse.*

Hegerl, E.J. 1984. Developing a conservation strategy for the mangrove ecosystems of Asia and Oceania, p.43-46. In: E. Soepadmo, A.N. Rao and D.J. Macintosh (eds.) *Proc. Asian Symp. Mangr. Environ. Res. Manage.* Univ. Malaya, Kuala Lampur, Malaysia.

Hodlgate, M.W., Bruce, J., Camacho, R.F., Desai, N., Mahtab, F.U., Mascarenhas, O., Maunder, W.J., Shihab, H. and Tewungwa, S. 1989. *Climate change : Meeting the challenge* (London : Commonwealth Secretariat, 1989), pp. ix, 131.

Houghton, R.A. 1990. Tropical deforestation and emissions of greenhouse gases. Paper from the Woods Hole Research Center, USA, presented to the Tropical forestry response Options Workshop of the Intergovernmental Group on Climate change, Sao Paulo, Brazil, January.

Houghton, J.T. 1991. Scientific Assessment of Climate Change : Summary of the IPCC Working Group I Report. pp.23-26. In: Jager, J. and H.L. Ferguson, Climate Change : Science, impacts and policy. *Proceedings of the second World Climate Conference.* Cambridge University.

IPCC. 1990. Policy makers summary of the scientific assessment of climate change. Report of Working Party I of the International Panel on Climate Change.

Isaksen, I.S.A. 1989. The beginnings of a problem in *Ozone Depletion : Health and Environmental Consequences*, Jones, R.R. and Wigley, T., Eds. (Chichester : John Wiley & Sons, 1989), pp.15-25.

Jayawardene, R.P. 1984. Mangroves of Asia and the Pacific : Present status and multiple practices : Sri Lanka country Report. Paper presented for Second Regional Task Force Meeting, Borog, Indonesia.

Jin-Eong, Ong. 1995. The ecology of mangrove conservation & management. *Hydrobiologia* **295**: 343-351.

Lugo, A.K. and Snedaker, S.C. 1974. The Ecology of Mangroves. *Ann. Rev. Ecology & Systematics* **5**: 39-64.

Lovelock, C.E., Clough, B.F. and Woodrow, I.E. 1992. Distribution and accumulation of Ultraviolet radiation absorbing compounds in leaves of tropical mangroves. *Planta* **188**: 143-154.

Macnae, W. 1966. Mangroves in eastern and southern Australia. *Aust. J. Bot.* **14**: 67-104.

———————— 1968. A general account of the fauna and flora of mangrove swamps and forests in the Indo-West Pacific region. *Adv. Mar. Biol.* **6**: 73-270.

Mall, L.P., Billore, S.K. and Amritphale, D. 1982. Certain ecological observations on mangroves of Andaman Islands. *Trop. Ecol.* **23**: 225-233.

Mathauda, G.S. 1957. The mangroves of India. *Proc. Mangr. Symp. Calcutta* 1957: 31.

Mcmillan, C. 1971. Environmental factors affecting seedling establishment of the black mangrove on the central Texas coast. *Ecology* **52**: 927-930.

Melillo, J.M., Callaghan, T.V., Woodward, F.I., Salati, E. and Sinha, S.K. 1990. Effects on ecosystems in climate change. *The IPCC Scientific Assessment*, Houghton, J.T., Jenkins, G.T. and Ephraums, J.J. Eds. (Cambridge : Cambridge University Press, 1990), pp.283-310.

——————, McGuire, A.D., Kicklighter, D.W., Moore, B., III, Vorosmarty, C.J. and Schloss, A.L. 1993. Global climate change and terrestrial net primary production. *Nature* **363**: 234-240.

Milliman, J.D., Broadus, J.M. and Gable, F. 1989. Environmental and economic implications of rising sea level and subsidising deltas : The Nile and Bengal examples. *Ambio* **18**: 340-345.

Mitchell, J.F.B., Manabe, S., Meleshko, V. and Tokioka, T. 1990. Equilibrium climate change and its implications for the future, in climate change. *The IPCC Scientific Assessment*, Houghton, J.T., Jenkins, G.J. and Ephraums, J.J. Eds. (Cambridge : Cambridge University Press, 1990) pp.131-172.

—————— and Gregory, J.M. 1992. Climatic consequences of emissions and a comparison of IS92a and SA90, in climate change 1992. *The supplementary Report to the IPCC scientific assessment*, Houghton, J.T., Callander, B.A. and Varney, S.K. Eds. (Cambridge University Press, 1992) pp.171-175.

Mongia, A.D., Dagar, J.C. and Bandyopadhyay, A.K. 1992. Potential of degraded mangrove areas of Andaman and Nicobar Islands for Agriculture and other alternative uses. *Agrochimica* **36**(4-5): 312-323.

Moore, R.T., Miller, P.C., Albright D. and Tiezen, L.L. 1972. Comparative gas exchange characteristics of the mangrove species during the winter. *Photosynthetica* **6**: 387-394.

————————, ————————, Ehleringer, J. and Lawrence, W. 1973. Seasonal trends in gas exchange characteristics of three mangrove species. *Photosynthetica* **7**: 387-394.

Moorman, F.R. and Pons, L.J. 1975. Characteristics of mangrove soils in relation to their agricultural land use and potential. *Proc. Intern. Mangr. Inst. Food and Agr. Sci.* Florida 1975: 520-547.

Muniyandi, K. and Natarajan, R. 1985. Incidence of seedling formation in *Rhizophora lamarckii* Montr. at Pichavaram Mangrove, Tamil Nadu, India. *Journ. Bomb. Nat. Hist. Soc.* **82**: 441-442.

Odum, E.P. 1971. Fundamentals of Ecology. *Philadelphia*: W.B. Saunders.

———————— 1978. Mangroves: Their value and perpetuation. *Nat. and Res.* **14**: 6-13.

Odum, W.E. and Heald. 1972. Trophic Analysis of an estuarine mangrove community. *Bull. Mar. Sci.* **22**: 671-738.

Othman, Muhammad Akhir, 1994. Values of mangroves in coastal protection. *Hydrobiologia* **285**: 277-282.

Passioura, J.B., Ball, M.C. and Knight, J.H. 1992. Mangroves may salinize the soil and in so doing limit their transpiration rate. *Funct. Ecol.* **6**: 476-481.

Percival, M. and Womersley, J.S. 1975. Floristics and ecology of the mangrove vegetation of Papua New Guinea. 96 pp. Papua New Guinea National Herbarium, Lae, Papua New Guinea.

Pernetta, J.C. and Osborne, P.L. 1990. Deltic flood plains : The mangroves of the Gulf of Papua and the Fly River, Papua New Guinea. In: *Implications of Expected Climate changes in the south Pacific Region : An overview*, Pernetta, J.C. and Hughes, P.J., Eds. UNEP Regional Seas Reports and Studies 128. pp.20-217.

———————— and Elder, D.L. 1992. Climate, sea level rise and the coastal zone : management and planning for global changes. *Ocean and Coastal Management*, **18**: 113-160.

——————— 1993. Mangrove forests, climate change and sea level rise : Hydrological influences on community structure and survival, with examples from the Indo-West Pacific. A Marine Conservation and Development Report. IUCN, Gland, Switzerland. vi + 46 pp.

Precht, H. 1967. A survey of experiments on resistance adaptation. In: *The Cell and Environmental Temperature* (A.S. Troshin, ed.) pp. 307-321. Pergamon Press, Oxford.

Rabanall, H.K. 1977. *Philippine Journal of Fisheries*, **14**(2): 191-212.

Rabinowitz, D. 1978. Dispersal properties of mangrove propagules. *Biotropica* **10**: 47-57.

Radhakrishna, K. 1978. Primary productivity of the Bay of Bengal. *Indian J. Mar. Sci.* **7**(1): 58-60.

Rajagopalan, M.S. 1987. Mangroves as component of coastal ecosystems of the Andamans. In: *Proc. Symp. Manag. Coastal Ecosystems and Oceanic Resources of Andamans.* Andaman Sci. Assoc. Port Blair pp.1-7.

Rajagopalan, V.R. 1952. Ecological adaptation of the mangrove vegetation of Pichavaram. *J. Annamalai Univ.* Annamalainagar, India.

Ramachandran Nair, Gopinathan, P.V. and Gopinathan, C.P. 1983. Primary production in coastal waters. *Bull. Cent. Mar. Fish Res. Inst.* **34**: 20-32.

Rao, A.N. 1987. Mangrove ecosystems of Asia and Pacific. p.1-48. In: Ricardo M. Umali *et al* (eds.). *Mangroves of Asia and the Pacific: status and management.* UNESCO. Quezon city, Metro Manila.

Roychowdhury, K.C. 1964. Prospects of better utilization of Goran and mangrove barks from Sunderbans for the tannin industry. Paper in *Symp. On Role of Forestry in the Development of Sunderbans.* 24-Parganas, India.

Ryther, J.H. 1969. Photosynthesis and fish production in the sea. *Science* **166**: 72-76.

Saenger, P.S., Hegeri, E.J. and Davie, D.S. (eds.) 1983. Global status of mangrove ecosystem. IUCN Commission of Ecology, Netherlands, 88 pp.

Sahni, K.C. 1957. Mangrove forests in the Andaman & Nicobar. Paper in *Mangrove Symp*. 16-19 Oct. 1957. Calcutta, India.

Scholander, P.F., Van Dam, L. and Scholander, S.I. 1955. Gas Exchange in the roots of mangroves. *Am. J. Bot*. **42**: 92-98.

——————, Hammel, H.T., Hemmingsen, E.A. and Garay, W. 1962. Salt balance in Mangroves. *Plant Physiol*. **37**: 722-729.

—————— 1968. How mangroves desalinate water. *Physical Plant* **21**: 251-261.

Scholl, D.W. 1963. Recent sedimentary record in mangrove swamps and rise in sea level over the southwestern coast of Florida : Part 1. *Marine Geology* **1**: 344-366.

—————— 1964. Recent sedimentary record in mangrove swamps and rise in sea level over the southwestern coast of Florida : Part 2. *Marine Geology* **2**: 343-364.

Sidhu, S.S. 1963. Studies on Mangroves. *Proc. Indian Acad. Sci*. **33**(8): 129-136.

Singh, N.T. and Mongia, A.D. 1985. Soils of the Andaman and Nicobar Islands. *J. Andaman Sci. Assoc*. **1**(1): 28-34.

——————, —————— and Ganeshamurthy, A.N. 1988. Soils of Andaman and Nicobar Islands. *CARI, Port Blair Bulletin*. no.1 pp.78.

Singh, V.P., Mall, L.P., George, A. and Pathak, S.M. 1987. A new record of some mangrove species from Andaman Islands and their distribution. *Ind. For*. **113**(3): 214. 2 figs.

Snedaker, S.C. and Pool, D.J. 1973. Mangrove forest types and biomass. pp.C-1-C-13. In: *The role of mangrove ecosystems in the maintenance of environmental quality and a high productivity of desirable fisheries.* Final report to Bureau of Sport fisheries and Wildlife. Contract No.14-16-008-606, Center for Aquatic Sciences.

——————— 1978. Mangroves : Their value and perpetuation. *Nat. and Res.* **14**: 6-13.

———————, Meeder, J.F., Ross, M.S. and Ford, R.G. 1994. Discussion of Ellison, J.C. and Stoddart, D.R., 1991. Mangrove ecosystem collapse during predicted sea level rise. Holocene analogues and implications. *Journal of Coastal Research* **7**(1): 151-165, *J. Coastal Res.* **10**: 497-498.

Spercely, A.P. 1977. The role of pneumatophores in sedimentary processes. *Marine Geol.* **24**: 31-37.

Stoddart, D.R. and Pethick, J.S. 1984. Environmental hazard and coastal reclamation : problems and prospects in Bangladesh. In: Bayliss Smith, T.P. and Wanmali, S. (eds.), *Understanding Green Revolutions : Agrarian Change and Development Planning in South Asia.* Cambridge University Press, pp.339-361.

Teas, H.J. 1979. Silviculture with saline water. *Biosaline concept* 1974: 117-161.

Tomilson, P.B. 1978. *Rhizophora* in Australia - some clarification of taxonomy and distribution. *J. Arnold Arbor.* **59**: 156-169.

——————— 1986. The Botany of Mangroves p.333, 340-342, figs.B.70. Cambridge University.

Titus, J.G., Leatherman, S.P., Everts, C.H., Kriebel, D.L. and Dean, R.G., 1985. Potential Impacts of sea level rise on the Beach at Ocean City, Maryland, EPA 230-1085-013. Washington, DC.: U.S. Environmetal Protection Agency.

Trexler, Mark C., Mintzer, Irving M. and Moomaw, William, R. 1990. Global Warming : an assessment of its scientific basis, its likely impacts and potential response strategies. Paper delivered at UNECE/USEPA Workshop on the Economics of Sustainable Development, Washington, DC. January.

United Nations Educational, Scientific and Cultural Organisation (UNESCO) 1973. International Classification and Mapping of Vegetation. UNESCO, Paris.

Van der Pijl, L. 1936. Fledermause and Blumen. *Flora* **31**: 1-40.

Vicente, V.P., Singh, N.C. and Botello, A.V. 1989. Ecological implications of potential climate change and sea level rise in the wider Caribbean. In: *Implications of climatic changes in the wider Caribbean Region*, Maul, G.A., Ed. UNEP (OCA)/CARWG.1/INF.3, pp.193-223.

Waheedkhan, M.A. 1957. Ecological studies on the mangrove forests in India. *Proc. Mangr. Symp*. Calcutta, 1957. pp.97-109.

Waisel, Y. 1972. Biology of Halophytes. New York : Academic Press.

Walsh, G.E. 1974. Mangroves : A review. pp.51-174. In: Reimhold, R.J. and Queen, W.H. (eds.) *Ecology of Halophytes*. Academic Press, New York.

Walter, Jorg and Robert, Ayres 1990. Global Warming : damages and costs. Paper presented to IPCC, January.

Ward, J.V. 1988. Palynology of Kosrae, Eastern Caroline Islands : recoveries from pollen rain and Holocene deposits. *Rev. Palaeobot. Palynot*. **55**: 247-271.

Warrick, R.A. and Oerlemans, H. 1990. Sea level rise in climate change. *The IPCC Scientific Assessment*, Houghton, J.T., Jenkins, G.J. and Ephraums, J.J., eds. (Cambridge : Cambridge University Press, 1990), pp.257-281.

Wigley, T.M.L. and Raper, S.C.B. 1992. Implications for climate and sea level of revised IPCC scenarios. *Nature* **357**: 293-300.

Womersley, J.S. 1983. An introduction to the nomenclature and taxonomy of the mangrove flora in Papua New Guinea and adjacent areas, p.87-90. In: H.J. Teas (ed.). *Biology and Ecology of Mangroves*. The Hague : Dr W. Junk Publ.

Yanney-Ewusie, J. 1980. Elements of Tropical Ecology. London : Heinemann Educ. Books.

INDEX